MEDICAL PHYSICS

*Introduction to*
*Digital Filtering*

# Introduction to Digital Filtering

*Edited by*

R. E. Bogner

*Department of Electrical Engineering*
*University of Adelaide, Australia*

*and*

A. G. Constantinides

*Department of Electrical Engineering*
*Imperial College of Science and Technology, London*

*A Wiley–Interscience Publication*

JOHN WILEY & SONS
London · New York · Sydney · Toronto

*Library of Congress Cataloging in Publication Data:*

Bogner, R. E.
Introduction to digital filtering.

"A Wiley–Interscience publication."
1. Digital filters (Mathematics) I. Constantinides, A. G., joint author. II. Title.

TK7872.F5B64      621.3815′32      74–4924
ISBN 0 471 08590 1

Printed in Great Britain by J. W. Arrowsmith Ltd.
Winterstoke Road, Bristol.

# Contributing Authors

P. F. BLACKMAN — *Department of Electrical Engineering, Imperial College of Science and Technology, Exhibition Road, London, S.W.7, England.*

R. E. BOGNER — *Electrical Engineering Department, University of Adelaide, Box 498 D, G.P.O. Adelaide, South Australia.*

G. C. BOWN — *The General Electric Company, Hirst Research Centre, Wembley, Middlesex, England.*

R. F. W. COATES — *School of Engineering Science, University College of North Wales, Bangor, Wales.*

A. G. CONSTANTINIDES — *Department of Electrical Engineering, Imperial College of Science and Technology, Exhibition Road, London, S.W.7., England.*

V. B. LAWRENCE — *Department of Electrical Engineering, Imperial College of Science and Technology, Exhibition Road, London, S.W.7., England.*

G. B. LOCKHART — *Department of Electrical Engineering, University of Leeds, Leeds, England.*

P. A. LYNN — *Department of Electrical Engineering, University of Bristol, Bristol, England.*

# Preface

The field of digital filtering and digital signal processing became important originally when the need arose to model analogue-signal-processing schemes on digital computers. The same area is now considered a distinct and separate discipline and as a result of this separation and treatment a very fertile and important area of research has been uncovered. The importance of digital filtering processing and digital filtering lies primarily in the immediate application of the techniques both in real-time digital computers or special purpose hardware and in off-line processing.

Owing to the rather accelerated growth of the subject a need was felt for post-experience courses and such courses were organized and given at Imperial College, London. The final product of these courses is represented by the contributions included in this book. More specifically we have arranged the material as follows:

Chapter 1 presents a general review of the motivations and achievements in the field of digital filters. The basic ideas of sampling $Z$-transform techniques are given in Chapter 2. These ideas form the basis on which the remainder of the book is supported. In Chapter 3 the general characteristics of digital filters are examined and the concepts are developed in the time and frequency domains. The use of existing analogue filters for designing digital filters is presented in Chapter 4 whereas Chapter 5 contains some procedures on the direct design of recursive digital filters. Non-recursive digital filters are examined in Chapter 6 and the concepts of Fourier transforms in Chapter 7. A useful design procedure employing sampling in the frequency domain is given in Chapter 8 and in Chapter 9 the same ideas are approached for filters with integer coefficients. The effects of rounding and quantization on the performance of different filters are examined in Chapter 10. Finally in Chapter 11 optimization techniques are examined which are useful from the design point of view.

Thanks are due to many contributors who made their expert knowledge available to us, and to our colleagues for their support.

R. E. BOGNER*
A. G. CONSTANTINIDES
Imperial College, London.

* R. E. Bogner is now with the University of Adelaide, Australia.

# Contents

## Chapter 1

# Introduction

*R. E. Bogner*

My purpose is to set a perspective and to suggest why we are where we are, doing what we are and where we expect to be going.

There is a diversity of opinion about what we should include in the term 'digital filtering'. It seems appropriate to include most signal processing systems in which the signals are represented by sequences of values, available only at discrete intervals of 'time'. This approach allows us to consider systems in which signal samples are stored as 'analogue' values on capacitors, or which are made up of lengths of transmission line—most of the mathematics is the same as that describing the purist's digital filter made of digital hardware.

### 1.1 History; trends

The accelerating development of digital computing power for commercial reasons has led to increasingly reliable, faster, smaller and better value hardware[1,2,3,4,5,6] (Figure 1.1).

Even before automatic digital computers, there were primitive, arithmetic-limited applications of related techniques, e.g.:

Fourier harmonic analysis, modification in amplitude and phase, component and summation.

Smoothing of time series by 'windows', i.e. convolution.

Correlation or regression analysis.

Periodogram analysis equivalent to comb filtering.

Autoregressive analysis[8], equivalent to recursive filtering.

These tasks were rapidly adaptable to digital computers. They were usually applied to what might be called signal analysis problems—finite, short duration 'signals', e.g. seismic records, tides, waves, thermal behaviour of buildings, economic cycle analysis. Automatic digital computers made it practical to increase the detail in such tasks and carry out also numerical convolution of long time series. Such tasks are equivalent to simulation of analogue signal processing. Improvements in speed and value-for-money led naturally to simulation of communication systems where the signals could be conveniently represented by samples. The complication and flexibility requirements of many speech processing experiments resulted in the emergence of audio and electroacoustics workers among the most enthusiastic promoters of computer processing of signals[9]. Advantages over analogue experimental methods are flexibility, reliability, precision and economy.

2

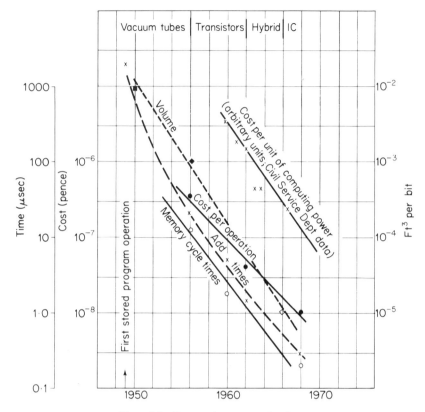

**Figure 1.1**　Progress in computer hardware

There is almost a continuum of in-between possibilities, from big, general-purpose number crunchers, through small general-purpose computers dedicated to signal processing tasks and programmed by software, to very restricted, specially designed equipment, possibly able to do specific tasks more efficiently or faster. The general-purpose machine has the expensive central processor available for many tasks—all the additions and multiplications being done by it, possibly for several filters, but the system may be limited in speed by this factor. The trend is toward very flexible semi-specialized processors which may perform tens of arithmetic operations simultaneously under programme control.[10,11,12].

We have the computer race to thank for the availability of high performance, small and economical digital hardware. Current technology provides many thousands of circuit elements per square centimetre[13], and we may soon expect to see available specialized integrated circuits for digital filter functions.

Digital systems offer the advantages over analogue circuits of being absolutely stable and reproducible, convenient for integrated circuit manufacture, of arbitrary flexibility and of compatibility with digital transmission techniques.

However, it is almost certain that there will always be a frontier where 'speed' is a limitation—at present this occurs at about one megahertz, but this is steadily advancing.

## 1.2 Comparison of continuous and digital filters—characteristic equations

In continuous, lumped systems we have elements which perform integration and differentiation:

$$v(t) = \frac{\int i(t)\,dt}{C} \qquad i(t) = \frac{\int v(t)\,dt}{L}$$

$$i(t) = C\frac{dv(t)}{dt} \qquad v(t) = L\frac{di(t)}{dt}$$

and also scaling devices—amplifiers, resistances, transformers:

$$v = Ri; \qquad v_2 = Av_1; \qquad v_2 = Nv_1.$$

The resultant equations are linear integro-differential equations, e.g. the first-order system of Figure 1.2(a) is described by the differential equation

$$\frac{L}{R}\frac{dv_2}{dt} + v_2 = v_1 \tag{1.1}$$

and the characteristic or impulse response is shown in Figure 1.2(b). The solutions of the equations are always of the form of exponentially damped (positively or negatively) sinusoids or cosinusoids characteristic of the system, plus forced responses characteristic of the input.

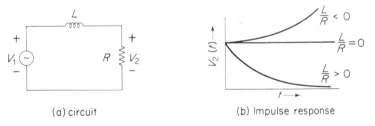

(a) circuit            (b) Impulse response

**Figure 1.2**  First-order continuous or analog system

Superposition applies perfectly if the components are perfect (linear).

The variables are defined at every instant of time.

In digital systems, variables are available only as values which may change at discrete values of time. The operations we have are addition and multiplication, and also delay by multiples of the time, $T$ seconds, between samples—the clock period or sampling interval. The delay comes about from the possibility of storing the values of the signal as long as we like.

The simplest example is the first-order system (Figure 1.3a). In this, the value of the output, $y(nT)$, is equal to $x(nT)$ delayed by one sampling interval, i.e.

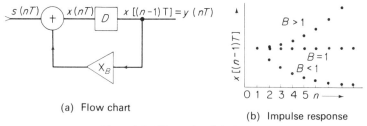

(a) Flow chart

(b) Impulse response

**Figure 1.3** First-order digital system

the previous value of $x$ is taken as the latest value of $y$. The system shown is described by the equation

$$x(nT) = s(nT) + Bx[(n - 1)T] \qquad (1.2)$$

$$\therefore \quad x[(n - 1)T] = \frac{x(nT)}{B} - \frac{s(nT)}{B}$$

$$\therefore \quad x(nT) - x[(n - 1)T] = x(nT) - \frac{x(nT)}{B} + \frac{s(nT)}{B}$$

$$\therefore \quad B\Delta_1 x(nT) + (1 - B)x(nT) = s(nT) \qquad (1.3)$$

where $\Delta_1$ is the first difference operator, such that

$$\Delta_1 x(nT) = x(nT) - x[(n - 1)T].$$

Equation (1.3) is the *difference* equation of the system, directly comparable with the differential equation (1.1) of the continuous system:

$$\frac{L}{R} \frac{dv_2}{dt} + v_2 = v_1.$$

(1.3) is a *difference equation*; and hence so is (1.2). Similarly, equations of discrete time delay systems are *difference equations*. These have the same role as differential equations in continuous systems. In both cases, the equations. are linear, a great blessing, allowing us to apply the principle of superposition.

The system shown has an impulse response (Figure 1.3b) which is exponential—like that of the $L$–$R$ circuit, but sampled. This result is easily seen for this simple system. Consider a unit pulse, $s(0) = 1$, applied when the value of $y$ is zero, followed by subsequent inputs $s(nT)$ of zero. At successive clock periods we have:

| $n$ | $s(nT)$ | $y(nT) = x[(n - 1)T]$ | $x(nT) = s(nT) + By(nT)$ |
|---|---|---|---|
| 0 | 1 | 0 | 1 |
| 1 | 0 | 1 | $0 + B$ |
| 2 | 0 | $B$ | $0 + B^2$ |
| ... | ... | ... | ... |
| $n$ | 0 | $B^{n-1}$ | $B^n$ |

As with continuous filters, corresponding damped sinusoidal oscillations are available as elemental responses of higher-order systems.

## 1.3 Time and frequency

As for continuous filters, it is usual to specify digital filters in time or in frequency. Specification in time is the description of the impulse response as a (usually finite) sequence of numbers. These numbers may be used directly via a convolution (because the systems obey superposition):

$$y(t) = \int_{-\infty}^{\infty} x(\lambda)h(t - \lambda)\,d\lambda;$$

$$y(nT) = \sum_{r} x(rT)h[(n - r)T]$$

or may be used via a transform method. The frequency specification—amplitude and phase—is probably the most common specification in both continuous and discrete filters. In digital filters, the discrete Fourier Transform (DFT) allows us to specify the frequency values directly, and Fast Fourier Transform (FFT) algorithms often make the use of transform methods advantageous. Often when a time specification is given, it turns out to be best to transform it to the frequency domain and use Fourier transforms:

$$
\begin{array}{c}
y(nT) \ \ = \ x(nT) \ \ * \ \ h(nT) \\
\text{Inverse} \ \uparrow \text{FT} \qquad \downarrow \text{FT} \qquad \downarrow \text{FT} \\
Y(mF) = X(mF) \times H(mF)
\end{array}
$$

For analytical work and conceptual purposes it is frequently useful to describe a transfer function by its poles and zeros—frequencies (usually complex) at which the transfer function becomes infinite or zero, e.g. in the continuous case the maximally flat or Butterworth filter (4th order)—Figure 1.4. All realizable linear systems are describable in these terms. Having decided that a particular pole–zero pattern provides a suitable characteristic, we have to design a system with the required pattern.

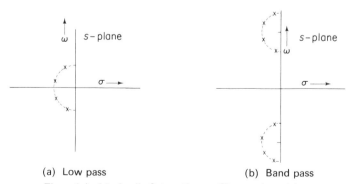

(a) Low pass       (b) Band pass

**Figure 1.4** Maximally flat continuous filter—pole positions

6

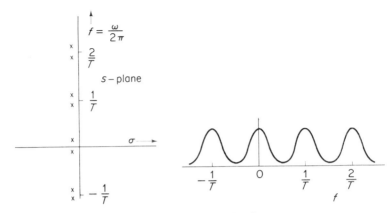

(a)  Repetition of poles and zeros          (b)  Frequency response

**Figure 1.5**  Periodicity in frequency: (a) repetition of poles and zeros; (b) frequency
response

One significant difference between continuous and digital systems arises
here—digital systems have $p$–$z$ patterns which are periodic in frequency
($\omega$) (Figure 1.5). The repetition is a result of the essential sampled representation
of the signals, and occurs at intervals of $1/T$ Hz where $T$ is the sampling interval.
It is found convenient to represent this cyclic behaviour by remapping the
$s$-plane using

$$z = e^{sT}$$

where the $\omega$ axis is found to transform into the unit circle, and each circuit of the
circle corresponds to one period along the $\omega$ axis.

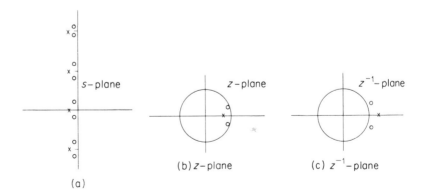

**Figure 1.6**  Mapping $s$ into $z$

We find that $z$ is a variable which has the usefulness for digital filters which
$s$ has in continuous systems. The 'z-transform' has a role corresponding to
the Laplace transform.

## 1.4 Imperfections

There are practical problems in each type—continuous filters have problems in the stability and accuracy of components; digital filters have absolutely precise components, but they can have only quantized values. This may not be significant in large computer simulations where we have floating point numbers of many significant figures, but can be very important where we are dealing with fixed point numbers in specially constructed hardware. The variables in continuous systems are specified with infinite precision; in digital systems they are quantized and every stage of addition or multiplication becomes a possible source of roundoff errors or noise. These factors influence systems architecture.

## 1.5 Hybrid systems

Most of the ideas concerning digital filters are applicable to other discrete-time filters, and we have the possibility of various hybrid systems—to economize on (say) central processor power. Thus we may use digital storage, analog multiplication and addition; or delay in discrete-time steps, but with analogue magnitudes, stored on capacitors.

## 1.6 Synthesis procedures

The basic synthesis strategies are similar in both cases—approximation to a desired characteristic by a mathematical function which is known to be realizable. The specification may be in time or frequency or ...? Because of the considerable body of experience available in continuous filter characteristics (e.g. Butterworth, Chebycheff, Elliptic, Bessel) it is sometimes convenient to make use of these, via appropriate transformations, for finding suitable difference equations. In contrast to continuous filter design, we find that the realization, once a suitable equation has been found, is often very easy. Practical factors like cost or noise are likely to affect the choice among many possible realizations.

As with many other areas of engineering design, automatic computer procedures for synthesis are making an impact in digital filters[14]. This could be considered a natural state of affairs—the designers of digital filters are likely to have the relevant familiarity with computer methods and, also, digital filters do not yet have an adequate body of analytical design procedures. They may not ever need to have! Automatic optimization can take into account simultaneously many factors such as non-analytical design criteria, quantizing effects and costs.

## 1.7 Why time and frequency?

It seems reasonable that a system whose behaviour is a function of time should be specified or described in time functions. But what relation do frequency and sinusoids have to flipflops, shift registers, etc? We have a long familiarity

8

with continuous lumped systems, whose differential equations yielded the widely useful sine, cosine and exponential. The difference equations describe discrete-time, continuous-amplitude systems and yield similar functions, sampled.

But we have discrete-amplitude systems under discussion—particularly if we want to save 'bits', i.e. use reduced resolutions. There seems to be a case for considering functions which are more closely similar to discrete system behaviour, e.g. Walsh functions[15,16] (Figure 1.7). These are orthogonal over the interval $-\frac{1}{2}, \frac{1}{2}$ and have many properties like sin and cos, e.g. there are Walsh Transforms which correspond to Fourier Transforms. So far, communication systems based on these principles have not been clearly superior to others. The challenge remains—conventional methods of specification are not necessarily the appropriate ones.

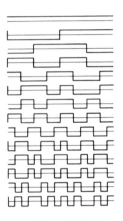

Figure 1.7    Some Walsh functions

What do we really want our filter to do? Is the criterion really 'flat to $x$ Hz then cut off to less than $-60$ db by $y$ Hz'? We should consider our signal processing engine as a whole, e.g. as a device which, given a waveform, makes a decision. In fact each output sample is simply a result of a number of calculations on previous inputs. Let them be the most profitable ones. We have the opportunity in digital systems of introducing time-variable characteristics or adaptive behaviour, at little cost beyond mental effort.

Summary of salient features

| Property | Continuous filter | Digital filter |
|---|---|---|
| Variables | Continuous in time<br>Continuous in magnitude | Discrete time 'Samples'<br>Usually discrete magnitudes |
| Mathematical operations | $d/dt$, $dt$, $Xk$, $\pm$ | Delay, $Xk$, $\pm$ |
| Characteristic equation | Linear differential | 'Linear' difference (occasionally 'logical-difference') |
| Characteristic responses | Damped sinusoids, cosinusoids | Samples of damped sinusoids, cosinusoids |
| Superposition, convolution | Yes | Yes |
| Transforms | Fourier<br>Laplace | Discrete Fourier<br>$z$ |
| Frequency domain | $s$ plane<br>$-\infty \leqslant \omega \leqslant \infty$ | $z$ plane, $z = e^{sT}$. Cyclic behaviour in $\omega$ corresponds to circuits of unit circle |
| *Imperfections:* | | |
| Cost | Static | Decreasing rapidly |
| Speed | To optical frequencies | To MHz |
| 'Components': | Initial tolerances<br>Drift<br>Non-linearity and over-loading | Coefficient rounding<br>Absolutely stable<br>Quantization and over-flow |
| Noise | Thermal, shot, etc. | Quantizing, aliasing; low level limit cycles |

## References

1. Klerer, M., and Korn, G. A., *Digital Computer Users' Handbook*, McGraw-Hill, 1967.
2. Grabbe, E. M., Ramo, S., and Wooldridge, D. E., *Handbook of Automation, Computation and Control*, Vol. 2, 1959.
3. Wilkes, M. V., *Automatic Digital Computers*, Methuen, 1956.
4. *I.E.E. Convention, 1956*, I.E.E. Publications.
5. Rösen, S., 'Electronic computers—a historical survey', *Computing Surveys*, 1 No. 1, (March 1969).
6. *Automatic Data Processing—The Next Ten Years* (Internal Report), Civil Service Dept.
7. Whitaker, E. T., and Robinson, G., *The Calculus of Observations*, Blackie, 1924.
8. Yule, 1927, quoted in Bartlett, *An Introduction to Stochastic Processes*, Cambridge University Press, 1961.
9. See, for example, the Special Issues of *Trans. IEEE Audio and Electroacoustics* devoted to Digital Filtering, Sept 1968 and June 1970.
10. Rayner, P. J. W., *A Hardware Digital Filter with Programming Facilities*, Symposium on Digital Filtering, Imperial College, July 1970.

11. *Programmable Digital Filters* (a new product bulletin), Rockland Systems Corporation, 1970.

12. Bergland, G. D., 'Fast Fourier transform hardware implementation—an overview', *Trans. IEEE Audio and Electroacoustics*, **AU-17** No. 2, 104–108 (June 1969).

13. Heath, F. G., 'Large-scale integration in electronics', *Sci. Am*, **222** No. 2, 22–31 (Feb. 1970).

14. Rabiner, L. R., Gold, B., and McGonegal, C. A., 'The approximation problem for nonrecursive digital filters'. [Probably in *Trans. IEEE Audio and Electroacoustics*, (1970).] Reported at IEEE Workshop on Digital Filtering, Harriman, N.J., Jan 1970.

15. Harmuth, H. F., 'A generalized concept of frequency and applications', *Trans. IEEE on Information Theory*, **IT-14** No. 3, 375–382 (1968).

16. Gibbs, J. E., and Millard, M. J., *Walsh Functions as Solutions of a Logical Differential Equation*, D.E.S. Report No. 1, Nat. Phys. Lab., Ministry of Technology, 1969.

# Chapter 2

# Introduction to Sampling and
# z-transforms

*P. F. Blackman*

## 2.1 Introduction

In many situations in communications, control or data handling it is required to transmit information available in continuous form by means of a series of samples. For example this may arise due to the need to use a single channel to carry signals for several separate information links. Individual samples may be transmitted as analogue samples or converted to binary form and converted back to analogue form after reception. Alternatively the binary representation may be processed in some way in its digital form. There is thus the need to analyse systems carrying sampled signals.

In the investigation of systems carrying continuous signals the Laplace transformation technique which converts signals from the time domain into an $s$-domain representation greatly simplifies analysis. In a similar way the analysis of systems carrying sampled signals is much simplified by the use of the $z$-transformation which converts sampled signals in the time domain to an algebraic representation in the $z$-domain.

This chapter gives a brief outline of the $z$-transform principles and their application to the analysis of sampled data systems.

## 2.2 Modulation

A carrier signal $\cos \omega_c t$ may be expressed in exponential form as

$$\cos \omega_c t = \frac{e^{+j\omega_c t} + e^{-j\omega_c t}}{2}$$

and represented on a frequency axis by two line spectrum components as in Figure 2.1(a). If the carrier is modulated by a signal $k \cos \omega_m t$, to give conventional amplitude modulation, Figure 2.1(b), the process may be represented in exponential form as

$$(1 + k \cos \omega_m t) \cos \omega_c t = \left[ 1 + \frac{k}{2}(e^{+j\omega_m t} + e^{-j\omega_m t}) \right] \frac{e^{+j\omega_c t}}{2}$$

$$+ \left[ 1 + \frac{k}{2}(e^{+j\omega_m t} + e^{-j\omega_m t}) \right] \frac{e^{-j\omega_c t}}{2} \qquad (2.1)$$

giving six frequency components

$$\pm \omega_c ; \qquad \pm \omega_c \pm \omega_m$$

12

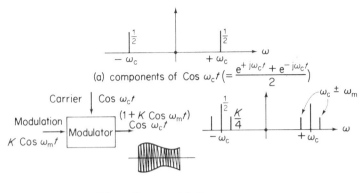

(a) components of $\cos \omega_c t \left( = \dfrac{e^{+j\omega_c t} + e^{-j\omega_c t}}{2} \right)$

(b) amplitude modulation

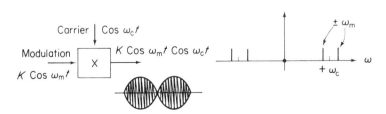

(c) amplitude modulation with several carriers

**Figure 2.1** Amplitude modulation: (a) components of $\cos \omega_c t [= (e^{+j\omega_c t} + e^{-j\omega_c t})/2]$; (b) amplitude modulation; (c) amplitude modulation with several carriers

in which the modulation components are present centred on the carrier components to give side-band signals. If the carrier contains a number of components, $\cos \omega_{c1} t$, $\cos \omega_{c2} t$, ..., the modulated carrier spectrum would contain the modulation components repeated about all the carrier components, see Figure 2.1(c).

If suppressed carrier modulation is used, the modulation process is a direct multiplication

$$k \cos \omega_m t \cos \omega_c t$$

and the modulated output contains the same side-band signals, but the carrier does not appear, Figure 2.2. Again if the carrier contains a number of components the side-band signals will be centred about each carrier frequency

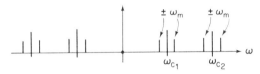

**Figure 2.2** Suppressed carrier modulation

value. The general properties of suppressed carrier modulation may be summarized as

(i) if the modulation is zero there is no output
(ii) the carrier can only be obtained in the output by applying a d.c. input.

For both forms of modulation there is no modulation component ($\pm\omega_m$) in the output.

## 2.3 Sampling

If a signal $f(t)$ is sampled by a switch closing periodically for a short time with sampling interval $T$, as in Figure 2.3(a), the output will be a train of finite width pulses. However for many analytic applications it is more convenient to assume an *ideal sampler*, which is considered to have the property of providing an output which is a train of impulses as in Figure 2.3(b), where the impulses have an area $f(nT)$ equal to the magnitude of $f(t)$ at the sampling instants $t = 0, T, 2T, \ldots, nT \ldots$. This process is often referred to as *impulse sampling*.

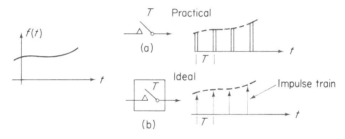

**Figure 2.3**   Practical and ideal sampling

Ideal sampling may be compared with suppressed carrier modulation in which the carrier is a continuous train of unit impulses as in Figure 2.4, which may be expressed as

$$f_c(t) = \sum_{n=-\infty}^{+\infty} \delta(t - nT) \tag{2.2}$$

and the output will be the product of the input, $f(t)$, and carrier signal

$$f(t)f_c(t) = \sum_{n=-\infty}^{+\infty} f(t)\delta(t - nT) = f^*(t) \tag{2.3}$$

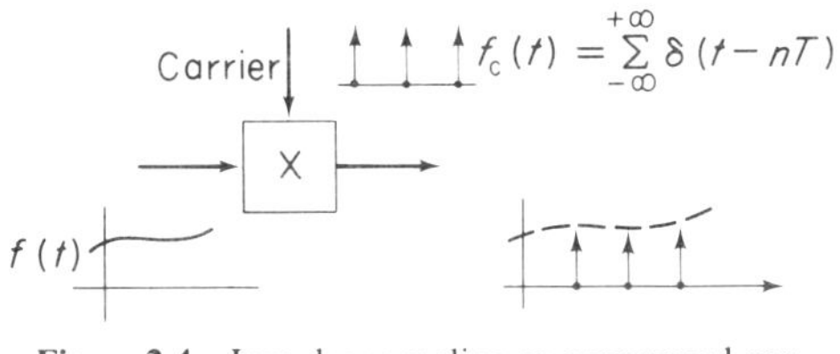

**Figure 2.4**   Impulse sampling as suppressed carrier modulation

sampled signals being represented by a star. If $f(t) = 0$, the modulation is zero and there is no output, and if $f(t)$ is constant the output is the carrier only. Thus impulse sampling has the same properties, (i) and (ii), as suppressed carrier modulation.

## 2.4 Laplace transform of a sampled signal

In lumped, linear, continuous (i.e. not sampled) systems, time functions often take the form of a sum of general exponential components

$$(t > 0), \qquad f(t) = a_1 e^{s_1 t} + a_2 e^{s_2 t} + \cdots = \sum_{k=1}^{k=m} a_k e^{s_k t} \qquad (2.4)$$

$$(t < 0), \qquad f(t) = 0$$

where $s_k$ are the poles of the transformed signal which will be either real or in conjugate pairs in the complex domain, and $e^{s_k t}$ are the natural modes.

If impulse sampling of a single component, $f_k(t) = a_k e^{s_k t}$, is considered as in Figure 2.5, the sampled component will be given by

$$f_k^*(t) = \sum_{n=0}^{n=\infty} a_k e^{s_k t} \delta(t - nT)$$

$$= a_k[\delta(t) + e^{s_k T}\delta(t - T) + e^{2 s_k T}\delta(t - 2T)\ldots] \qquad (2.5)$$

which represents a series of delayed impulses in which the coefficients $a_k e^{s_k T}$, $a_k e^{2 s_k T}, \ldots$ are complex constants.

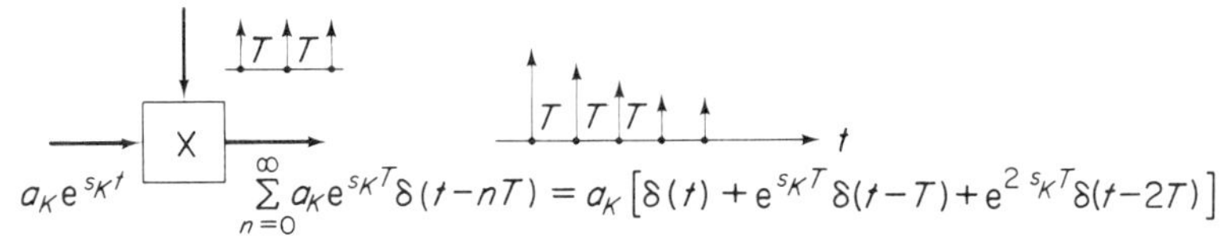

$$a_K e^{s_K t} \quad \boxed{X} \quad \sum_{n=0}^{\infty} a_K e^{s_K T}\delta(t-nT) = a_K\left[\delta(t) + e^{s_K T}\delta(t-T) + e^{2 s_K T}\delta(t-2T)\right]$$

**Figure 2.5**   Impulse sampling of a single exponential component

Since the Laplace transform of a delayed impulse is given by

$$\mathscr{L}\delta(t - nT) = e^{-snT}$$

the transform of the impulse sampled component is given by

$$\mathscr{L}f_k^*(t) = F_k^*(s) = a_k[1 + e^{s_k T} e^{-sT} + e^{2 s_k T} e^{-2sT} + \cdots] \qquad (2.6)$$

which may be written in closed form as

$$F_k^*(s) = \frac{a_k}{(1 - e^{s_k T} e^{-sT})} = \frac{a_k}{(1 - e^{(s_k - s)T})} \qquad (2.7)$$

for $|e^{(s_k-s)T}| < 1$. This result should be compared with the transform of the corresponding continuous signal

$$\mathcal{L} a_k e^{s_k t} = \frac{a_k}{(s - s_k)}. \tag{2.8}$$

### 2.5 Complex plane representation of sampled signals

The Laplace transform of a general sampled exponential component

$$F_k^*(s) = \frac{a_k}{1 - e^{(s_k - s)T}} \tag{2.9}$$

has poles where

$$e^{(s_k - s)T} = 1 \tag{2.10}$$

which correspond with

$$(s_k - s)T = 0, \qquad \text{i.e. } s = s_k \tag{2.11}$$

or

$$(s_k - s)T = \pm jn2\pi, \qquad \text{i.e. } s = s_k + \frac{jn2\pi}{T}, \qquad -\infty < n < \infty. \tag{2.12}$$

In the $s$-plane, see Figure 2.6, these results give a pole at $s_k$ corresponding with the continuous signal, and a line of additional poles spaced at intervals $2\pi/T$. The spacing $2\pi/T$ corresponds with the *sampling frequency* $\Omega$ given by

$$\Omega T = 2\pi. \tag{2.13}$$

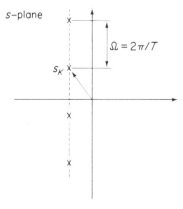

**Figure 2.6** Pole pattern for single sampled exponential

Since $F_k^*(s)$ has an infinite number of poles, it could be represented as a summation of individual poles with the general form

$$F_k^*(s) = \sum_{n=-\infty}^{+\infty} \frac{b_n}{(s - s_n)} \tag{2.14}$$

where $s_n$ are the poles, and the individual coefficients $b_n$ are the residues at the poles. For the general ratio of polynomials

$$F(s) = \frac{N(s)}{D(s)} \tag{2.15}$$

the residue at a simple pole $s_j$ is given by the relation

$$\text{residue} = \frac{N(s_j)}{\dfrac{d}{ds}D(s)\bigg|_{s=s_j}}. \tag{2.16}$$

Applying this result to $F_k^*(s)$ gives

$$\text{residue} = \frac{a_k}{T\,e^{(s_k - s)T}}\bigg|_{s = s_k + jn2\pi/T = s_k + jn\Omega} \qquad (-\infty < n < \infty)$$

$$= \frac{a_k}{T} \quad \text{for all } n. \tag{2.17}$$

hence

$$F_k^*(s) = \frac{a_k}{T}\left[ \cdots \frac{1}{s - (s_k - jn\Omega)} \cdots + \frac{1}{s - (s_k - j\Omega)} + \frac{1}{(s - s_k)} \right.$$

$$\left. + \frac{1}{s - (s_k + j\Omega)} \cdots \frac{1}{s - (s_k + jn\Omega)} \cdots \right]. \tag{2.18}$$

For the complete continuous signal

$$f(t) = a_1\,e^{s_1 t} + a_2\,e^{s_2 t} + \cdots a_k\,e^{s_k t} \tag{2.19}$$

the transformed expression is

$$F(s) = \frac{a_1}{(s - s_1)} + \frac{a_2}{(s - s_2)} + \cdots \frac{a_k}{(s - s_k)} \tag{2.20}$$

and the individual terms may be combined together over a common denominator to give

$$F(s) = \frac{c(s - s_a)(s - s_b)\cdots}{(s - s_1)(s - s_2)\cdots} \tag{2.21}$$

where the numerator gives the zeros, $s = s_a$, $s_b \ldots$ of $F(s)$, as in Figure 2.7. These are the $s$-values for which the sum of the individual components vanishes. Finally, the corresponding frequency spectrum $F(j\omega)$ can be obtained by

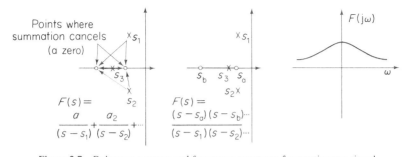

**Figure 2.7** Pole-zero pattern and frequency spectrum for continuous signal

evaluating $F(s)$, in either sum or product form, along the imaginary axis in the
$s$-plane.

If the same signal $f(t)$ is now sampled, the corresponding transformed expression becomes

$$F^*(s) = \frac{a_1}{1 - e^{(s_1 - s)T}} + \frac{a_2}{1 - e^{(s_2 - s)T}} + \cdots \frac{a_k}{1 - e^{(s_k - s)T}} \tag{2.22}$$

and in the $s$-plane each individual term represents a line of poles with spacing
$j\Omega$. Hence the poles of $F^*(s)$ are the same pattern as for $F(s)$, but repeated at
intervals $j\Omega$ as in Figure 2.8a. Finally the individual terms in $F^*(s)$ could be
expressed as a series to give the general form

$$
\begin{aligned}
F^*(s) = {}& \frac{a_1}{T}\left[ \cdots \frac{1}{s - (s_1 - j\Omega)} + \frac{1}{(s - s_1)} + \frac{1}{s - (s_1 + j\Omega)} + \cdots \right] \\
& + \frac{a_2}{T}\left[ \cdots \frac{1}{s - (s_2 - j\Omega)} + \frac{1}{(s - s_2)} + \frac{1}{s - (s_2 + j\Omega)} + \cdots \right] \\
& + \frac{a_k}{T}\left[ \cdots \frac{1}{s - (s_k - j\Omega)} + \frac{1}{(s - s_k)} + \frac{1}{s - (s_k + j\Omega)} + \cdots \right] \\
= {}& \frac{1}{T}[\cdots F(s - j\Omega) \quad + F(s) \quad + F(s + j\Omega) \ldots] \\
= {}& \frac{1}{T}\sum_{n=-\infty}^{+\infty} F(s + jn\Omega). \tag{2.23}
\end{aligned}
$$

Thus the effect of sampling is to repeat the pole–zero pattern of the original
signal at intervals $j\Omega$ in the $s$-plane, the overall result being the sum of the
individual patterns. Hence the frequency spectrum along the imaginary axis
is the sum of the original spectrum $F(j\omega)$ displaced by intervals $j\Omega$ as in Figure
2.8(b). When the repeated pole–zero patterns are added together there will be
poles at the same points as in the individual patterns, but zeros now occur at
new locations. These locations are the points where the sum of the repeated
patterns is zero, which are not the same points at which the individual patterns
have zeros.

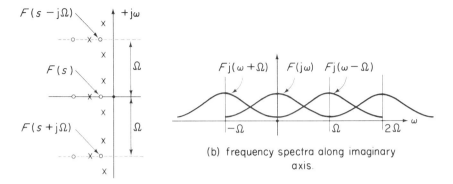

(a) pole–zero pattern for
sampled signal.

(b) frequency spectra along imaginary
axis.

**Figure 2.8**  Pole-zero patterns and frequency spectrum for sampled signal: (a) pole-zero
pattern for sampled signal; (b) frequency spectra along imaginary axis

## 2.6   Signal recovery; the sampling theorem

Since in general for a sampled signal the spectra overlap, the total signal
present at any one frequency $\omega_j$, see Figure 2.9, depends on the addition of
contributions from a number of spectra which must be added with due regard
to the individual phase angles leading to a tedious calculation. If only a few
spectra are considered in order to simplify the analysis, there may be doubt
about the accuracy of the result. For this reason frequency response methods
do not have the breadth of application in sampled-data systems that they have
in continuous systems. Increasing the sampling rate ($\Omega$) causes the spectra to
move further apart reducing the amount of overlap.

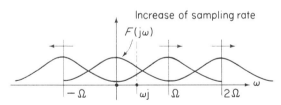

**Figure 2.9**   Signal present at a particular frequency

It can also be seen in Figure 2.9 that the spectrum of the original signal $F(j\omega)$
is present. This represents a difference compared with normal modulation
processes in which the original signal is not present in the modulated output.
Hence in principle the original signal could be recovered by filtering. Clearly
the original signal can only be recovered perfectly if the following conditions
are satisfied:

(i) the spectra do not overlap
(ii) the filter has an ideal *brickwall* characteristic and hence passes $F(j\omega)$
perfectly, and completely eliminates all other spectra.

Condition (i) requires that $F(j\omega)$ has no components at frequencies exceeding $\Omega/2$, hence the spectra do not overlap, see Figure 2.10(a). Condition (ii) cannot normally be met since any practical filter does not have the ideal characteristic of Figure 2.10(b), but always modifies $F(j\omega)$ slightly.

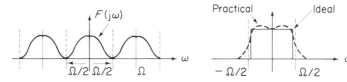

(a) condition for possible signal recovery          (b) filter characteristic required

**Figure 2.10**  Signal recovery and the sampling theorem: (a) condition for possible signal recovery; (b) filter characteristic required

These results show that in theory it is possible to recover a band-limited signal perfectly from samples, provided that the sampling frequency $\Omega$ exceeds $2\times$ (highest frequency component in the signal). This is essentially the *Sampling Theorem*. Also if a system which is low-pass is driven by a signal obtained from a relatively high speed sampler, the effect of sampling can be ignored.

### 2.7   Relation between s-plane and z-plane: z-transforms

As previously established (Section 2.3) the Laplace transform of a sampled general exponential signal $e^{s_k t}$, is given by

$$F^*_{(s)} = \frac{1}{1 - e^{-sT} e^{s_k T}} \tag{2.24}$$

and the term $e^{-sT}$ leads to an infinity of poles in the s-plane. In order to ease the problem of interpretation and analysis it is convenient to make a substitution

$$e^{sT} = z, \quad \text{or } e^{-sT} = z^{-1} \tag{2.25}$$

which leads to a corresponding $F(z)$, a *z-transform*, and to investigate this in the z-plane. It should be noted that $e^{-sT}$ corresponds with a delay of one sampling interval $T$ in the s-domain, while $z^{-1}$ corresponds to a similar delay in the z-domain. Since

$$z = e^{sT} = e^{(\alpha + j\omega)T} \tag{2.26}$$

this gives

$$|z| = e^{\alpha T}, \quad \angle z = \omega T \tag{2.27}$$

so that a point $s_k$ in the s-plane transforms to a point $z_k$ in the z-plane as in Figure 2.11(a).

Considering the angle of $z_k$ gives that

$$-\pi < \angle z_k < +\pi \quad \text{for } -\pi < \omega T < \pi, \quad \text{or } -\Omega/2 < \omega < \Omega/2 \tag{2.28}$$

20

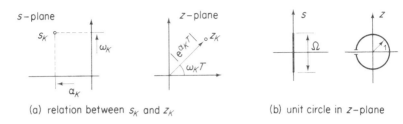

(a) relation between $s_K$ and $z_K$        (b) unit circle in $z$-plane

**Figure 2.11** Relation between $s$-plane and $z$-plane: (a) relation between $s_k$ and $z_k$; (b) unit circle in $z$-plane

and since

$$|z| = e^{\alpha_k T} \qquad (2.29)$$

the path along the imaginary axis in the $s$-plane in Figure 2.11(b) transforms into the *unit circle* in the $z$-plane, since $\alpha = 0$ along the imaginary axis and hence $|z| = 1$. Thus a strip $\Omega$ wide in the $s$-plane transforms to cover the entire $z$-plane, and successive strips in the $s$-plane transform into the same $z$-plane. Strictly the strips transform into successive Riemann surfaces superimposed on the $z$-plane. The general relation is indicated in Figure 2.12(a), the left half of the $s$-plane strip, which is the stable region in the $s$-plane, being transformed inside the unit circle in the $z$-plane.

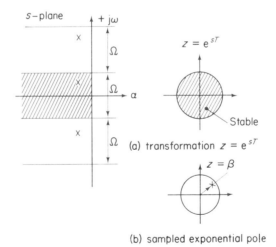

(a) transformation $z = e^{sT}$

(b) sampled exponential pole

**Figure 2.12** Transformation from $s$- to $z$-plane: (a) transformation $z = e^{sT}$; (b) sampled exponential pole

The important effect of the transformation is that since the poles of the transform in the $s$-plane are spaced along a vertical line at intervals j$\Omega$, all poles transform to a single pole in the $z$-plane. Hence the $z$-transform of a general sampled exponential has only a single pole in the $z$-plane.

Thus the Laplace transform of a sampled exponential $e^{s_kt}$ given by

$$F^*(s) = \frac{1}{1 - \beta e^{-sT}} \tag{2.30}$$

where $\beta = e^{s_kT}$, has the corresponding $z$-transform

$$F(z) = \frac{1}{1 - \beta z^{-1}} = \frac{z}{z - \beta} \tag{2.31}$$

giving a pole in the $z$-plane at $z = \beta$, as in Figure 2.12(b), and will be inside the unit circle if $|e^{s_kT}| < 1$, corresponding with a stable exponential.

The analytic steps in obtaining the $z$-transform of a general continuous exponential signal

$$f(t) = e^{s_kt} \qquad (t > 0, \, f(t) = 0, \, t < 0) \tag{2.32}$$

are summarized below.

(i) Sampling,

$$f^*(t) = \delta(t) + e^{s_kT}\delta(t - T) + e^{2s_kT}\delta(t - 2T) + \cdots. \tag{2.33}$$

(ii) Laplace transformation,

$$F^*(s) = 1 + e^{s_kT} e^{-sT} + e^{2s_kT - 2sT} \cdots. \tag{2.34}$$

(iii) Substitution $(z = e^{sT})$:

$$F(z) = 1 + e^{s_kT}z^{-1} + e^{2s_kT}z^{-2} + \cdots \tag{2.35}$$

$$= \frac{1}{1 - e^{s_kT}z^{-1}} = \frac{z}{z - e^{s_kT}} \tag{2.36}$$

which has a pole at $z = e^{s_kT}$. It is important to note that the location of the $z$-plane pole changes as the sampling interval $T$ is altered.

## 2.8  General relation between signals and pole locations

Considering the general exponential signal $e^{s_kT} = e^{(\alpha_k + j\omega_k)T}$, the angle turned through in one sampling interval is given by $\omega_k T$, which is also the angle of the corresponding pole in the $z$-plane, see Figures 2.11(a), 2.13(a). The radius to the pole is given by $e^{\alpha_k T}$ which is the change in magnitude of the exponential during one sampling interval.

From these results it is easy to correlate the general relation between pole locations in the $z$-plane and the corresponding exponentials, general relations being indicated in Figure 2.13(b). Poles inside the unit circle, having $|z| < 1$, correspond with decaying signals and vice-versa. A pole on the positive real axis, $\angle z = 0°$, corresponds with samples from a real exponential. A pole on the negative real axis corresponds with samples from an oscillatory signal with 2-sample intervals per cycle. A signal with 4-sample intervals per cycle corresponds with a pole on the imaginary axis. For oscillatory signals the actual sampling instant during the cycle is related to a numerator term or 'zeros' of the transform.

22

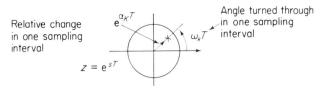

(a) pole location and sampling interval

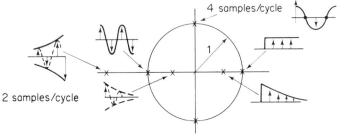

(b) pole locations and corresponding sampled signals

**Figure 2.13** Pole and time function correlation: (a) pole location and sampling interval; (b) pole locations and corresponding sampled signals

Some examples showing the variation of zeros with sampling instant are given below, and the corresponding pole-zero patterns shown in Figure 2.14:

(i) $f(t) = \cos t, (\omega = 1)$, sampled at 4 per cycle, i.e. $T = \pi/2$

$$f(t) = \frac{e^{jt} + e^{-jt}}{2} \to F(z) = \frac{1}{2}\left[ \frac{1}{1 - e^{j\pi/2}z^{-1}} + \frac{1}{1 + e^{-j\pi/2}z^{-1}} \right]$$

$$-\frac{z^2}{(z-j)(z+j)} \tag{2.37}$$

(ii) $f(t) = \sin t = \dfrac{e^{j} - e^{-jt}}{2j} \to F(z) = \dfrac{z}{(z-j)(z+j)}$ $\tag{2.38}$

(iii) $f(t) = (\cos t + \sin t) \to F(z) = \dfrac{z(z+1)}{(z-j)(z+j)}$ $\tag{2.39}$

In the inversion of $z$-transforms. zeros in $F(z)$ modify the residues at the poles and this changes the relation between the sampling instants and the continuous time signal.

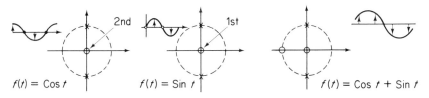

**Figure 2.14** Zero locations for various time functions

## 2.9 Inversion of z-transforms

There are three possible methods of inverting $z$-transforms to obtain the corresponding time function:

   (i) partial fraction expansion
   (ii) division
   (iii) contour integration and time patterns.

(i) *partial fraction expansion*
If an $F(z)$ can be expanded in partial fraction terms of the form corresponding with a general exponential, then the corresponding time function can be written directly, if

$$F(z) = \sum^k \frac{a_k}{1 - \beta_k z^{-1}} = \sum^k \frac{a_k z}{z - \beta_k} \tag{2.40}$$

then

$$f^*(t) = \sum^k a_k e^{s_k nT}, \qquad \text{where } e^{s_k T} = \beta_k \tag{2.41}$$

for example, see Figure 2.15(a)

$$F(z) = \frac{0 \cdot 5z}{(z^2 - 1.5z + 0.5)} = \frac{0 \cdot 5z}{(z - 1)(z - 0 \cdot 5)} = \frac{z}{(z - 1)} - \frac{z}{(z - 0 \cdot 5)} \tag{2.42}$$

and

$$f^*(t) = \left(1 - \sum_{n=0}^{\infty} (0 \cdot 5)^n \right) \delta(t - nT), \tag{2.43}$$

which can be expressed in exponential form as

$$f^*(t) = \left(1 - \sum_{n=0}^{\infty} e^{-0 \cdot 69n} \right) \delta(t - nT) \tag{2.44}$$

since $e^{-0 \cdot 69} = 0 \cdot 5$. This corresponds with the sampled step response of a time constant, where the time constant response changes by 50 per cent during each sample interval $T$, see Figure 2.15(a).

(ii) *division*
The partial fraction method assumes that $F(z)$ can be factorized easily which may not be the case, particularly for high order ratios of polynomials. In this

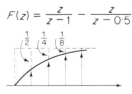

$$F(z) = \frac{z}{z-1} - \frac{z}{z-0 \cdot 5}$$

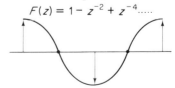

$$F(z) = 1 - z^{-2} + z^{-4} \ldots$$

(a) sampled step response                     (b) sampled cosine

**Figure 2.15**   $z$-transform inversion: (a) sampled step response: (b) sampled cosine

case the polynomial can be divided out to give a general series, and $f^*(t)$ can then be written directly in series form

$$F(z) = \frac{a_n z^n + a_{(n-1)}z^{n-1} + \cdots}{b_m z^m + b_{(m-1)}z^{(m-1)} + \cdots} = c_{(m-n)}z^{(n-m)} + c_{(m+1-n)}z^{(n-(m+1))}\cdots \quad (2.45)$$

$$f^*(t) = c_{(m-n)}\delta(t + (n-m)T) + c_{(m+1-n)}\delta(t + (n-m-1)T) + \cdots \quad (2.46)$$

For a realizable signal only zero or negative powers of $z$ can exist on the right-hand side, hence it is necessary that $n \leqslant m$ i.e. the number of zeros must not exceed the number of poles, and the difference $(n - m)$ is equal to the number of sample intervals elapsing before a sample value is obtained.

As an example, see Figure 2.15(b), if

$$f(t) = \cos \pi t / T \quad (2.47)$$

then

$$F(z) = \frac{z^2}{z^2 + 1} \quad (2.48)$$

which can be divided out to give

$$F(z) = 1 - z^{-2} + z^{-4} - z^{-6}\ldots \quad (2.49)$$

hence

$$f^*(t) = \delta(t) - \delta(t - 2T) + \delta(t - 4T) - \delta(t - 6T)\ldots \quad (2.50)$$

(iii) *contour integration*

As mentioned in (ii) a $z$-transform can be divided out to yield a series expansion in reciprocal powers of $z$

$$F(z) = c_0 z^0 + c_1 z^{-1} + c_2 z^{-2} + \cdots \quad (2.51)$$

where the coefficients $c_n$ are the values of $f(nT)$, and the series represents the summation of all orders of pole at the origin in the $z$-plane. In order to obtain the value of a particular coefficient, say $c_n$, both sides are multiplied by $z^{(n-1)}$ to give

$$z^{(n-1)}F(z) = c_0 z^{(n-1)} + c_1 z^{(n-2)} + \cdots c_n z^{-1} + c_{(n+1)}z^0 + \cdots \quad (2.52)$$

and then a contour integration carried out round a path enclosing all the poles to give

$$\oint z^{(n-1)}F(z)\,dz = \oint [c_0 z^{n-1} + \cdots c_n z^{-1} + \cdots]\,dz \quad (2.53)$$

In view of the general integral relation

$$\oint k z^n\,dz = 0, \qquad n \neq -1$$

$$= 2\pi jk, \qquad n = -1. \quad (2.54)$$

the only term that survives on the right-hand side is $c_n$. This leads to the general result that

$$c_n = f(nT) = \frac{1}{2\pi j} \oint z^{(n-1)} F(z)\, dz \qquad (2.55)$$

enabling the sample values $c_n$ to be obtained as a contour integral with the appropriate value of $n$. Such an integration would normally be evaluated by the method of residues to give

$$f(nT) = \sum \text{residues of } [z^{(n-1)} F(z)], \qquad \text{for } n = 0, 1, 2 \ldots \qquad (2.56)$$

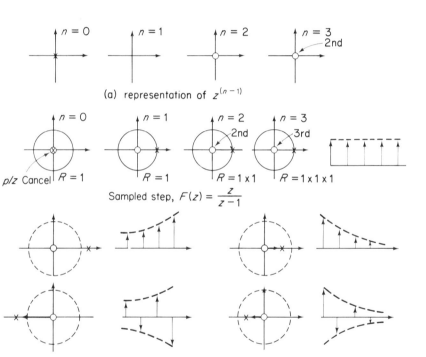

(a) representation of $z^{(n-1)}$

Sampled step, $F(z) = \dfrac{z}{z-1}$

(b) time pattern examples

(c) phasor representation of residues

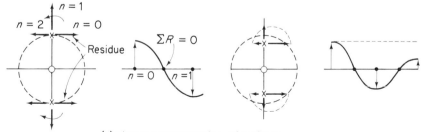

**Figure 2.16** Time patterns and residues: (a) representation of $z^{(n-1)}$; (b) time pattern examples; (c) phasor representation of residues

For a general function of $z$

$$F(z) = \frac{(z - z_a)(z - z_b)\ldots}{(z - z_1)(z - z_2)\ldots(z - z_k)\ldots} \qquad (2.57)$$

where $z_a$, $z_b$ ... are zeros, and $z_1$, $z_2$ ... are simple poles, the residue at a pole $z_k$ is given by

(residue at $z_k$) $= R_k = (z - z_k)F(z)$ evaluated for $z = z_k$.

The value of $R_k$ is in fact the value that the remainder of the function $F(z)$ gives at $z_k$, when the pole at $z_k$ is cancelled.

In determining $c_n$ in (2.55) the term $z^{(n-1)}$ corresponds to the introduction of an additional pole at the origin ($n = 0$), followed by increasing order zeros with increasing $n$ as shown in Figure 2.16(a). The residues at the poles of the original $F(z)$ may be obtained by calculation or by direct measurement on the plane, and the increasing order of zero at the origin multiples each residue by $z_k^{(n-1)}$. The idea just developed gives rise to the concept of *time patterns* which are useful to indicate many qualitative results. Since the residue at the pole is multiplied by $z_k$, the pole location, for each interval of time $T$, it follows that any pole outside the unit circle represents an increasing (unstable) component, and any pole inside the unit circle is stable, corresponding with a signal that dies away. Some examples are given in Figure 2.16(b).

The residue can be represented by a phasor drawn on the plane at the pole location, and the angle of the phasor changes by an angle $\angle z_k$ after each sampling interval this being due to the additional zero at the origin. The total value of $f(nT)$ is given by the sum of the residue phasors, hence only the real components need be considered, since the imaginary components cancel. Some examples are given in Figure 2.16(c)

## 2.10 Difference equations, $z$-transforms and transfer functions

Continuous time functions $f(t)$ can be considered as the solutions of differential equations which can be represented in analogue-computer-block-diagram form. In a similar manner an impulse train $f(nT)\delta(t - nT)$ can be considered as the solution of an appropriate difference equation, which can also be represented in block-diagram form.

In differential equation solution an essential operation is integration, represented in the $s$-domain by the relation $1/s$; in difference equations the essential operation is the unit delay represented by $1/z$ or $z^{-1}$. The $s$-domain consideration of differential equations leads to the development of transfer functions and the $s$-domain representation of time functions (Laplace transforms) and similarly the $z$-domain consideration of difference equations leads to the development of $z$-transfer functions and $z$-transforms.

The system of Figure 2.17(a) represents a delay $(z^{-1})$ with positive feedback $\beta$. There are two possible output points, $x_1$, $x_2$, and an input $u$, and it is considered

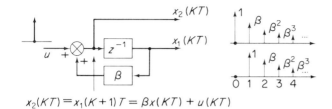

$$x_2(KT) = x_1(K+1)T = \beta x(KT) + u(KT)$$

(a) impulse response of feedback system

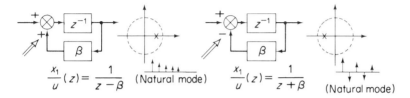

(b) effect of feedback sign reversal

**Figure 2.17** $z$-domain feedback system: (a) impulse response of feedback system; (b) effect of feedback sign reversal

that impulses can circulate in the system and that the delay time is $T$. If at $t = kT$, the impulse at $x_1$ is $x_1(kT)$, the input to the delay is $x_1(k+1)T$, since this reaches $x_1$ at $t = (k+1)T$. The signal fed back to the input is $\beta x_1$, and hence a difference equation can be written

$$x_1(k+1)T = \beta x_1(kT) + u(kT). \tag{2.58}$$

If a unit impulse is applied at $u$ at $t = 0$, this will appear at $x_2$ at $t = 0$, and will appear at $x_1$ at $t = T$, which will give a signal at $x_2, x_2(T) - \beta$. The impulse will continue to circulate round the loop being changed in amplitude by an amount $\beta$ for each circulation, leading to the impulse trains shown at $x_1$ and $x_2$.

The impulse trains could be represented in the $z$-domain as

$$X_2(z) = 1 + \beta z^{-1} + \beta^2 z^{-2} + \cdots = \frac{1}{1 - \beta z^{-1}} = \frac{z}{z - \beta} \tag{2.59}$$

$$X_1(z) = 0 + z^{-1} + \beta z^{-2} + \beta^2 z^{-3} \qquad = \frac{1}{z - \beta}. \tag{2.60}$$

The original difference equation may be expressed in the $z$-domain as

$$zX_1(z) = \beta X_1(z) + U(z) \tag{2.61}$$

or

$$(z - \beta)X_1(z) = U(z) \tag{2.62}$$

which leads to the transfer function relations between input and output of

$$\frac{X_1(z)}{U(z)} = \frac{1}{z - \beta} \quad \text{and} \quad \frac{X_2(z)}{U(z)} = \frac{z}{z - \beta}. \tag{2.63}$$

In each case it can be seen that the transfer function has the same denominator, which has a pole at $z = \beta$, which corresponds with a sampled exponential decay. Also the $z$-transform of the impulse response is the same as the transfer function. Thus there is the same relation between impulse response and transfer functions in the $s$- and $z$-domains, and in both domains systems have natural modes which are excited by an impulse, or any other input signal.

It should be noted that reversing the sign of the feedback as in Figure 2.17(b), does not cause instability, but merely moves the pole to $z = -\beta$, giving an oscillatory response, stability is determined by $|\beta|$, and for $|\beta| > 1$ the pole is outside the unit circle and the system is unstable.

In the case of a general input

$$U(z) = u_0 + u_1 z^{-1} + u_2 z^{-2} \ldots \tag{2.64}$$

each individual term in the input excites the impulse response of the system, so that if the transform of the impulse response of the system is given by

$$H(z) = h_0 + h_1 z^{-1} + h_2 z^{-2} + \cdots \tag{2.65}$$

as in Figure 2.18, the corresponding output will be

$$X(z) = u_0(h_0 + h_1 z^{-1} + h_2 z^{-2} + \cdots)$$
$$+ u_1 z^{-1}(h_0 + h_1 z^{-1} + h_2 z^{-2} + \cdots)$$
$$+ u_2 z^{-2}(h_0 + h_1 z^{-1} + h_2 z^{-2} + \cdots) \tag{2.66}$$

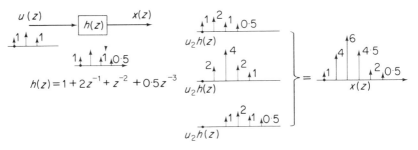

**Figure 2.18**  System response to general input

which can be seen to be given by the product of the input train and the impulse response, i.e.

$$X(z) = U(z)H(z) \tag{2.67}$$

or

$$\frac{X(z)}{U(z)} = H(z) \tag{2.68}$$

so that the $z$-transform of the impulse response is the $z$-transform function.

Hence $z$-transforms and transfer functions can be multiplied directly to yield overall results, but care must be taken when working from the continuous time domain because $z$-transform technique operates on the basis that the only signals are impulses, and there are no continuous signals in the system.

As an example Figure 2.19(a) shows the continuous step response of a time constant, and the Laplace transform of the output is given by the product of transforms of input and impulse response

$$X(s) = \frac{1}{s} \frac{1}{(1 + s\tau)} \tag{2.69}$$

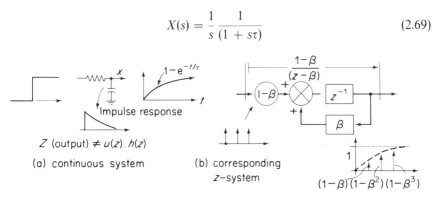

**Figure 2.19** Continuous and sampled systems: (a) continuous system; (b) corresponding $z$-system

but the expression for the $z$-transform of the output is *not*

$$X(z) = \frac{z}{z - 1} \cdot \frac{z}{z - \beta} = \frac{z^2}{(z - 1)(z - \beta)}: \quad \text{(wrong)}. \tag{2.70}$$

In order to obtain the $z$-transform it is necessary to convert the system response into the time domain and then to express the individual components in $z$-transform form. Hence

$$x(t) = 1 - e^{-t/\tau} \tag{2.71}$$

and

$$X(z) = \frac{z}{z - 1} - \frac{z}{z - \beta} = \frac{z(1 - \beta)}{(z - 1)(z - \beta)}: \quad \text{(correct)} \tag{2.72}$$

and the block diagram representation is given in Figure 2.19(b).

### References

The following contain sections on $z$-transformations and sampled data systems.

1. Kuo, B. C., *Automatic Control Systems*, Prentice-Hall, 1967.
2. Gupta, S. C., and Hasdorf, L., *Fundamentals of Automatic Control*, Wiley, 1970.
3. Gabel, R. A., and Roberts, R. A., *Signals and Linear Systems*, Wiley, 1973.

More detailed treatment
1. Lindorf, D. P., *Theory of Sampled Data Control Systems*, Wiley, 1965.
2. Kuo, B. C., *Discrete Data Control Systems*, Prentice-Hall, 1970.
3. Freeman, H., *Discrete Time Systems*, Wiley, 1965.
4. Jury, E. I., *Theory and Application of the z-Transform Method*, Wiley, 1964.

**Examples on z-transforms**

1. (a) Obtain the z-transform for $f(t) = e^{-t}$, with sampling interval $T = 1$, 0·5, 0·2 secs, and plot the pole for the transform in the z-plane. Write out the initial three or four terms in the series expansion of each transform and plot the relative magnitudes of the terms against a common time scale.

   (b) Obtain the z-transform for $f(t) = 1 - e^{-t}$ in product form

   $$\left[ \text{i.e. } \frac{az^{-1}}{(1 - bz^{-1})(1 - cz^{-1})} \right]$$

   with sampling interval $T = 1$ sec and plot the pole–zero pattern in the z-plane.

2. Obtain the z-transform for $f(t) = e^{-0·2t} \cos t$, for a general sampling interval $T$. [Hint: express $f(t)$ in exponential form and then re-combine individual z-transforms.]

   Plot the corresponding pole–zero pattern in the z-plane for $T = \pi/4, \pi/2$. Finally decide the general form of the pattern as $T \to \pi$ and the exact form when $T = \pi$.

3. For the system shown obtain the transfer function from input to $x_1$ and $x_2$, and write the first few terms of the series expansion of the transfer function. (Note: $1/(1 - a) = 1 + a + a^2 + a^3 + \cdots$.)

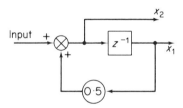

   If a single impulse is applied to the input determine subsequent signal values at $x_1$ and $x_2$ by considering the signals circulating in the system, and compare the results with those obtained in the first part of the question.

4. If the two outputs of the system in Q.3 are combined together obtain the

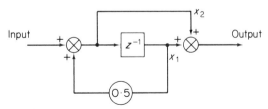

   series for the overall impulse response by adding the individual impulse responses at $x_1$ and $x_2$. Also write the overall transfer function and plot the pole–zero pattern in the z-plane.

Finally consider that an input of alternating unit impulses is applied i.e.

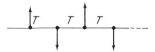

and obtain the output impulse train by summing appropriate delayed versions of the impulse response obtained in the first part of the question.

Compare the result with that obtained in Q.3, and also consider the result as the product of the transfer function and the $z$-transform of the input pulse train.

*Chapter 3*

# General Characteristics of Digital Filters

*A. G. Constantinides*

## 3.1 General comments

In linear classical network theory we have circuit elements which are characterized by linear mathematical operations on currents and voltages describing, as they do, their electrical properties. Thus we have for the resistor, inductor and capacitor the following mathematical formulae

$$v(t) = R \cdot i(t), \qquad v(t) = L\frac{di(t)}{dt}, \qquad i(t) = \frac{dv(t)}{dt}$$

where the symbols have their usual meaning. When these formulations are used in conjunction with Kirchhoff's laws, a set of *linear differential equations* (or in general, equations reducible to differential equations) are obtained which are characteristic of the particular linear network. However, the elements which are used in digital filters do not behave in the same way as the resistors, inductors and capacitors mentioned above. They are described in terms of an input–output relationship rather than a voltage–current relationship.

$$\text{Multiplier} \quad v_k = \alpha_i u_k$$
$$\text{Adder} \quad v(k) = u_1(k) + u_2(k) + \cdots + u_n(k)$$
$$\text{Delay} \quad v_k = u_{k-1}.$$

The significance of the various symbols and the representation of the operations is shown in Figure 3.1. Let us assume that digital filter elements are connected in any desirable way. In view of the operation of the digital filter elements their connection implies that the signals undergo a series of scaling operations in

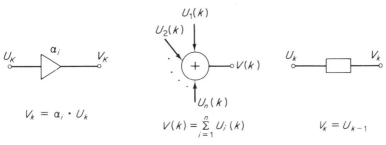

**Figure 3.1** Digital filter elements

33

amplitude, or are added to other signals or are delayed. Therefore, for a single input single output connection of these elements a *linear difference equation* will be obtained describing the input–output behaviour of the system.

To illustrate the above points consider the following example.

**Example 3.1.** A discrete-time system operates as follows. Each sample value of the output signal is made up by adding a corresponding input sample to a fraction $\alpha$ of the previous (in time) output sample.

*Solution.* The equation describing the operation above is derived as follows. Let the $k$th output sample be labelled $v_k$ so that the immediately previous output sample is $v_{k-1}$. Furthermore, let $u_k$ be the $k$th input sample value. The operation of the system is described by the relationship

$$v_k = u_k + \alpha v_{k-1} \tag{3.1}$$

which holds for all positive values of time (i.e. for all positive $k$). This is a linear first order difference equation. It is instructive to view the operation of this simple system from the time domain. For this purpose consider the input signal as given by

$$u_k = \{1, 0, 0, 0, \ldots\}.$$

The input signal is taken to be zero everywhere except at zero time when it has a value of unity. If the delay element of Figure 3.2 is assumed to have had no samples stored before the input is applied, then the operation of the system will be as follows. At zero time the signal at the input has a value of unity and since there is nothing stored in the delay, the output from it will be zero. The adder, therefore, will produce at its output a sample of unity value. This sample is the first sample of the output signal and is stored instantaneously in the delay element owing to its appearance at its input. The system will stay at this state until the next sample appears at its input after a time $T$. At this point in time the delay releases instantaneously the stored sample, which is of unity value, to appear at the point C and also simultaneously at the point D multiplied by a factor $\alpha$. Therefore, the next output sample appearing at B will be

$$v_1 = u_1 + \alpha v_0$$

$$= 0 + \alpha \cdot 1$$

$$= \alpha.$$

This value is the value of the sample stored in the delay element. In a similar manner the output sample at the following sampling instants will be given by,

$$v_k = \alpha^k, \qquad k = 0, 1, 2, 3. \tag{3.2}$$

In Figure 3.2 the output is drawn for three different positive values of $\alpha$. It is immediately apparent from this and also from equation (3.2) that the output samples decrease in magnitude with increasing time (i.e. increasing $k$) only

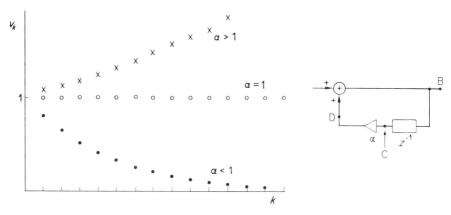

Figure 3.2   Time response of $v_k = u_k + \alpha v_{k-1}$ for $u_k = \{1, 0, 0, 0, \ldots\}$

when the modulus of the multiplier $\alpha$ is less than unity. Hence the system is stable if and only if $|\alpha| < 1$.

The output signal above can be considered as the *impulse response* of the system. This response has the familiar shape of a decaying exponential associated with a resistor–capacitor voltage decay.

### 3.2   The digital filter equation

From the above considerations and from the previous chapter the characterization of digital filters can be generalized as follows. Let $u_k$ be the $k$th sample value of the input signal of the digital filter. Then $u_{k-r}$ is the sample value of the same signal at the sampling instant $(k - r)T$. Similarly, if $v_k$ is the $k$th sample value of the output signal of the digital filter then $v_{k-r}$ is the sample value at the $(k - r)T$ sampling instant. Let it be required to compute the value of the $k$th output sample $v_k$. This computation, that is the operation of the digital filter, is carried out by combining linearly past output samples with past (and present) input samples. Such an operation is exemplified by the earlier example. Thus, if $(n + 1)$ input samples are used in the required linear combination process, along with $m$ previous output samples, the $k$th sample value of the output signal will be given in the form,

$$v_k = a_0 u_k + a_1 u_{k-1} + \cdots + a_n u_{k-n} - (b_1 v_{k-1} + \cdots + b_m v_{k-m}) \qquad (3.3)$$

where the coefficients are real and constant.

### 3.3   The transfer function of a digital filter

From above it is seen that a linear difference equation of the form given by equation (3.3) characterizes a linear digital filter operation. For transfer function considerations the discrete system transfer function is defined in terms of the

$z$-transform that was examined earlier. It is defined as the ratio of the $z$-transform of the output signal to the $z$-transform of the input signal. In the case of the example considered earlier the transfer function will be given by the expression

$$\frac{V(z)}{U(z)} = 1/(1 - \alpha z^{-1}).$$

**Example 3.2.** In a certain system the input and output signals have sample values as shown below. What is the transfer function of the system?

Input: 1, 0, 1, 0, 1, 0, 1, ...

Output: 1, 1, 1, 1, 1, 1, 1, ....

The $z$-transform of the input signal follows directly from the definition as,

$$U(z) = u_0 + u_1 z^{-1} + u_2 z^{-2} + \cdots$$
$$= 1 + 0 \cdot z^{-1} + 1 \cdot z^{-2} + \cdots$$
$$= 1 + z^{-2} + z^{-4} + \cdots$$
$$= \frac{1}{1 - z^{-2}} \quad \text{for } |z^{-1}| < 1.$$

Similarly, the $z$-transform of the output signal is given by,

$$V(z) = v_0 + v_1 z^{-1} + v_2 z^{-2} + \cdots$$
$$= 1 + z^{-1} + z^{-2} + \cdots$$
$$= \frac{1}{1 - z^{-1}} \quad \text{for } |z^{-1}| < 1.$$

Hence the transfer function of the discrete-time system is given by

$$G(z) = \frac{V(z)}{U(z)}$$
$$= (1/(1 - z^{-1}))/(1/(1 - z^{-2}))$$
$$= 1 + z^{-1}.$$

In general therefore the transfer function of equation (3.3) is given by

$$G(z) = \frac{V(z)}{U(z)} = \frac{a_0 + a_1 z^{-1} + \cdots + a_n z^{-n}}{1 + b_1 z^{-1} + \cdots + b_m z^{-m}} \tag{3.4}$$

where $V(z)$ is the $z$-transform of the output signal and $U(z)$ is the $z$-transform of the input signal. Equation (3.4) describes the transfer function of a general linear digital filter. The transfer function possesses a numerator and denominator which are in general non-zero. Based on the form of the transfer function, a classification is usually made as follows.

*Recursive digital filters* are those filters which possess a transfer function as given by equation (3.4) and with all common factors cancelled, the denominator coefficients are identically non-zero.

*Non-recursive digital filters*, however possess a transfer function which is a polynomial in $z^{-1}$ when all common factors are cancelled in equation (3.4). In this case, the transfer function is of the form,

$$G(z) = h_0 + h_1 z^{-1} + \cdots + h_n z^{-n}.$$

It was observed in Section 3.1 with reference to Figure 3.2 that the choice of a range of values for $\alpha$ in the transfer function $1/(1 - \alpha z^{-1})$ produced responses which either increased or decreased with time. This implies that there exists a region in which $\alpha$ is not allowed to be, otherwise the response increases indefinitely. In general it can be said of equation (3.4) that the denominator coefficients must be chosen in such a way that the transient response of the system does not increase without bound. This statement is intimately connected with the location of the *poles* of the system.

### 3.4  The poles and zeros of the transfer function

The numerator and denominator of the general transfer function of equation (3.4) are polynomials in the complex variable $z^{-1}$ with real coefficients, and as such they can be factorized in the form below.

$$G(z) = A\frac{(z^{-1} - \alpha_1)(z^{-1} - \alpha_2)\cdots(z^{-1} - \alpha_n)}{(1 - \beta_1 z^{-1})(1 - \beta_2 z^{-1})\cdots(1 - \beta_n z^{-1})}$$

where $\alpha_i$ and $\beta_i$ are either real or complex, but if they are complex then they occur in conjugate pairs. The factor A is real and constant.

The quantities $\{\alpha_i\}$, $i = 1, 2, 3, \ldots, n$ are called the *zeros* whereas $\{\beta_r^{-1}\}$, $r = 1, 2, 3, \ldots, m$ are called the *poles* of the transfer function. The zeros and poles are located on the $z^{-1}$ *plane*.

### 3.5  Stability

A digital filter is said to be stable when its poles have moduli which are greater than unity, which is equivalent to saying that the poles must be situated *outside* the circle $|z^{-1}| = 1$. (This circle is called the *unit circle*.)

For example, the digital filter whose transfer function is given by

$$G(z) = \frac{1 + z^{-1}}{(1 + 0.5z^{-1})(1 - 0.4z^{-1})}$$

is stable. It has two poles, one at $-1/0.5 = -2$ and one at $1/0.4 = 2.5$, i.e. both lie outside the unit circle on the $z^{-1}$ plane. However, the transfer function

$$G(z) = \frac{1 + z^{-1}}{(1 + 5z^{-1})(1 - 4z^{-1})}$$

is unstable. It has two poles, one at $-1/5 = -0.2$ and one at $1/4 = 0.25$, i.e. both lie inside the unit circle on the $z^{-1}$ plane.

## 3.6 Frequency response

To obtain the frequency response of a system we use as input a sinusoidal waveform and observe the output. If the system is linear, then the output will be the same sinusoidal waveform altered only in phase and amplitude. Since phase and amplitude are connected through complex quantities, one usually employs a complex sinusoid rather than a real one as input which is modified by the system in the sense that it is multiplied by a complex quantity depending on the frequency of the complex sinusoidal input. These ideas will be employed here to obtain the frequency response of the general digital filter as defined by equation (3.3).

Let the input be the complex sinusoid $\{e^{j\omega kT}\}$, $k = 0, 1, 2, \ldots$ Then, since the system is linear, the output can only be in the form $\{H(\omega) . e^{j\omega kT}\}$, $k = 0$, $1, 2, \ldots$, where $H(\omega)$ is a complex quantity that depends on $\omega$ only. Hence one can write in equation (3.3), $u_k = e^{j\omega kT}$ and $v_k = H(\omega) e^{j\omega kT}$ so that the difference equation becomes,

$$H(\omega) e^{j\omega kT} = a_0 e^{j\omega kT} + a_1 e^{j\omega(k-1)T} + \cdots + a_n e^{j\omega(k-n)T} - (b_1 H(\omega) . e^{j\omega kT} +$$
$$+ \cdots + b_m H(\omega) e^{j\omega(k-m)T}) \tag{3.5}$$

or, equivalently,

$$H(\omega) . e^{jk\omega T} . (1 + b_1 e^{-j\omega T} + \cdots + b_m e^{-jm\omega T}) = e^{j\omega kT} . (a_0 + a_1 e^{-j\omega T}$$
$$+ \cdots + a_n e^{-jn\omega T})$$

and hence by cancelling the common term $e^{j\omega kT}$ from both sides of the above equation one obtains,

$$H(\omega) = \frac{a_0 + a_1 e^{-j\omega T} + \cdots + a_n e^{-jn\omega T}}{1 + b_1 e^{-j\omega T} + \cdots + b_m e^{-jm\omega T}} \tag{3.6}$$

which is the frequency response of the system in complex form. The modulus of equation (3.6) yields the amplitude characteristic and the argument of $H(\omega)$ gives the phase characteristic of the system of equation (3.3).

Note that equation (3.6) is identical to the transfer function of equation (3.4) with $z^{-1}$ replaced by $e^{-j\omega T}$ which establishes the nature of $z^{-1}$ and the usefulness of the transfer function.

## 3.7 Realization structures for digital filter transfer functions

In this section the signal-flow diagrams are given for digital filter transfer functions in terms of the digital filter elements namely, the adder, multiplier and delay. These diagrams are also known as *realization structures* because it is in one of these forms that the practical realization is usually carried out.

The simplest form of realization is obtained by examining the general transfer function of equation (3.4). The introduction of an auxiliary $z$-transform $X(z)$ enables the transfer function $G(z)$ to be written in the form

$$G(z) = \frac{V(z)}{X(z)} \cdot \frac{X(z)}{U(z)} = \frac{1}{1 + b_1 z^{-1} + \cdots + b_m z^{-m}}$$
$$\times (a_0 + a_1 z^{-1} + \cdots + a_n z^{-n}). \qquad (3.7)$$

Equation (3.7) now can be partitioned into two transfer functions as follows

$$G_1(z) = a_0 + a_1 z^{-1} + \cdots + a_n z^{-n} = \frac{X(z)}{U(z)}$$

$$G_2(z) = \frac{1}{1 + b_1 z^{-1} + \cdots + b_m z^{-m}} = \frac{V(z)}{X(z)}.$$

The overall transfer function $G(z)$ is obtained by cascading the two realization structures of Figure 3.3(a) and Figure 3.3(b) which, after combining the three

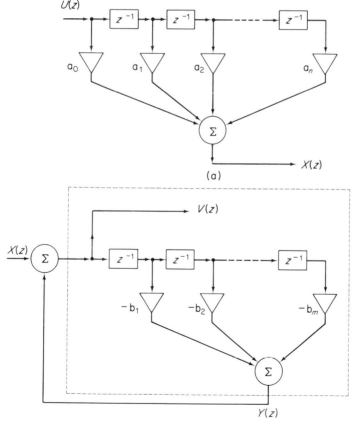

**Figure 3.3** (a) The Realization of $G_1(z) = a_0 + a_1 z^{-1} + \cdots + a_n z^{-n}$; (b) the Realization of $G_2(z) = 1/(1 + b_1 z^{-1} + \cdots + b_m z^{-m})$

adders, the final result is obtained as indicated in Figure 3.4. This realization is by no means unique since one can find other structures exhibiting the same algebraic transfer function. The different structures that realize a given transfer function are classified into *canonic* and *non-canonic* realizations. By the term *canonic realization* it is meant that *the number of delay elements employed is precisely equal to the order of the transfer function (i.e. the highest degree between the numerator and denominator polynomials).*

The followlng subsections deal with several canonic realizations which are commonly in use.

### 3.7.1 The series realization

The structure derived previously and shown in Figure 3.4 requires $(m + n)$ delays and hence according to the above definition it is non-canonic.

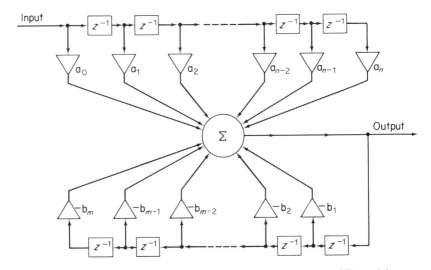

**Figure 3.4** Result of combining the two realizations structures of Figure 3.3

A rearrangement, however, of the structure can be effected as shown in Figure 3.5 where $n$ is assumed greater than $m$. This structure can be shown to have the required transfer function by straightforward analysis.

A particularly useful sub-structure, corresponds to the case when $n = m = 2$ as shown in Figure 3.6. This structure is examined below.

### 3.7.2 The biquadratic form

The sub-structure shown in Figure 3.6 forms the basic building block for the development of further and more complicated canonic structures. This form is known as the *biquadratic* and since its denominator is a quadratic it follows that a pair of real poles or complex conjugate poles are directly realizable by this structure. In the subsequent sub-sections the biquadratic structure is used to realize higher-order transfer functions in canonic forms.

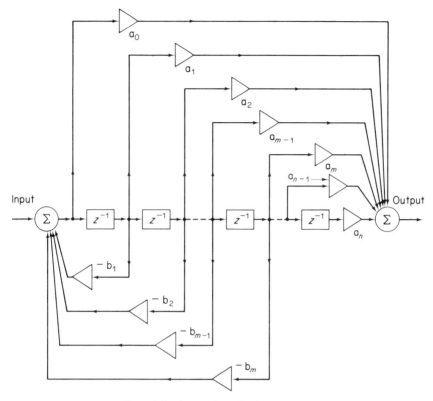

**Figure 3.5**  A canonic realization structure

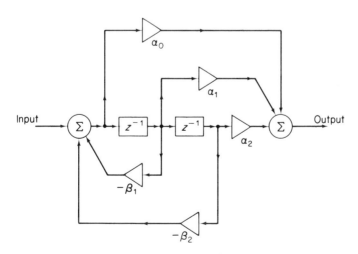

**Figure 3.6**  The biquadratic section

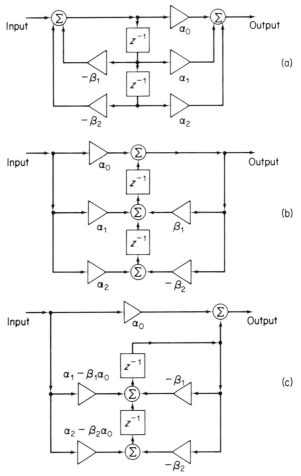

**Figure 3.7** Some canonic realizations of the biquadratic section

It should be pointed out that the canonic form of the biquadratic transfer function of Figure 3.6 is by no means unique. For example in Figure 3.7 we give other canonic forms which have identical transfer functions.

### 3.7.3 The cascade realization

The general transfer function of equation (3.4) can be factorized in the following form,

$$G(z) = \left[ \frac{\alpha_{01} + \alpha_{11}z^{-1} + \alpha_{21}z^{-2}}{1 + \beta_{11}z^{-1} + \beta_{21}z^{-2}} \right] \cdots \left[ \frac{\alpha_{0n} + \alpha_{1n}z^{-1} + a_{2n}z^{-2}}{1 + \beta_{1n}z^{-1} + \beta_{2n}z^{-2}} \right] \quad (3.8)$$

where all coefficients are real and constant and that a first order zero or pole are obtained by setting to zero the coefficients of the square terms in the appropriate factor. By selecting a numerator quadratic with a denominator quadratic and treating each factor of equation (3.8) as an individual transfer function one can consider the realization of equation (3.8) above as a cascade arrangement of biquadratics. This is shown in Figure 3.8 and is called the *cascade realization* which is a canonic form.

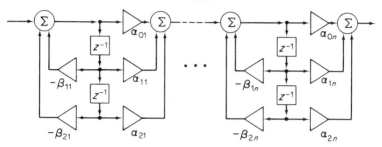

**Figure 3.8** The cascade realization

### 3.7.4 Parallel realization

Another canonic realization can be achieved when the transfer function of equation (3.9) is expanded in partial fraction form as shown below

$$G(z) = \gamma_0 + \sum_{i=1}^{n} \frac{\gamma_{0i} + \gamma_{1i}z^{-1}}{1 + \beta_{1i}z^{-1} + \beta_{2i}z^{-2}}$$

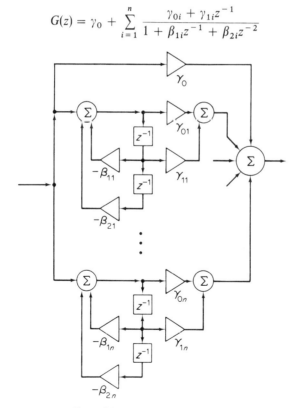

**Figure 3.9** The parallel realization

where $\gamma_0$, $\gamma_{0i}$ and $\gamma_{1i}$, $i = 1, 2, 3, \ldots, n$, are the appropriate real constant coefficients. The realization in this case is shown in Figure 3.9. Further structures are shown in Figure 3.10.

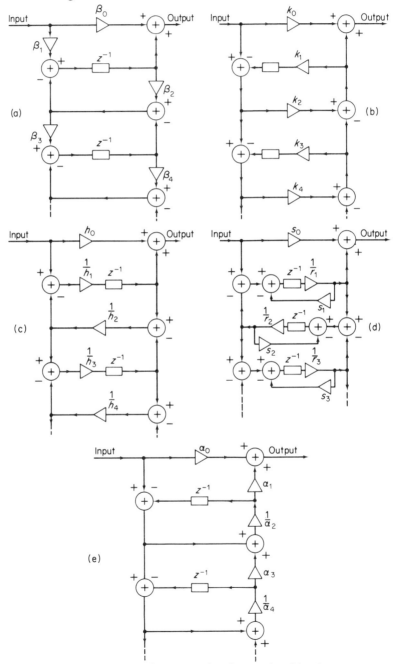

**Figure 3.10** Canonic structures based on continued fractions

# References

1. Rader, C. M., and Gold, B., *Digital Processing of Signals*, McGraw-Hill, 1969.

*Chapter 4*

# Synthesis of Digital Filters from Continuous Filter Data

*A. G. Constantinides*

**Summary**

The bilinear transformation for synthetizing digital filters has been described by various authors. The purpose of this lecture is to reexamine this transformation, modify it and propose a general multiband transformation from which the simpler transformations can be derived. It is shown that the Rader and Gold transformation is a special case of the proposed general transformation.

The technique of synthetizing digital filters by means of continuous to digital filter transformations, is akin to the s-plane standard techniques for continuous filters and therefore it is familiar to most engineers. The major advantage of the proposed methods, however, rests on the fact that digital filters can be synthetized by using tables of continuous s-plane filters.

## 4.1  Introduction

The development and use of digital filters followed, historically, that of continuous filters, and since there existed a wealth of literature describing synthesis methods for continuous filters, it was natural to resort to these when the construction of a digital filter was called for. A particular approach to the synthesis of digital filters has been to replace the integrating operator $1/s$, in a given continuous filter transfer function, by a numerical formula and higher orders of $1/s$ by higher order approximations.

A wealth of numerical integration formulae can be obtained from the literature on Numerical Analysis. For example the Gregory–Newton formula, the Simpson's rules, the trapezoidal approximation, etc., may be used effectively for the numerical representation of $1/s$.

In the field of digital filters (i.e. filters that are dealing with sampled signals, etc.) the most important approximation has been the trapezoidal one, not because of the 'closeness' of approximation but because of its inherent mapping properties.

Thus Kaiser[1] and Golden and Kaiser[2] suggest that instead of approximating the continuous integrating operator $1/s$ by a numerical formula, it may be more profitable to synthetize digital filters by transforming the continuous transfer function onto the z-plane.

The *modus operandi* of the bilinear transformation as given by Kaiser and used by several other authors[3] is given in Section 4.3. It is considered as a

47

mapping between the $s$-plane and the $z^{-1}$-plane and on this basis the general transformations are derived.

## 4.2 Indirect synthesis of digital filters

Statement of the problem.

'Given a transfer function of a continuous filter (i.e. a real rational function in $s$) to obtain the transfer function of a digital filter (i.e. a real rational function in $z^{-1}$) by mapping the complex variable $s$ through a functional relationship to the digital filter complex variable $z^{-1}$.'

Certain points need to be clarified before proceeding to the bilinear transformation. We shall assume that the complex variable $s$ is in the form $s = \Sigma + j\Omega$ and $z = \exp(\sigma T + j\omega T)$ because of the confusion that may arise when dealing with frequencies on the $s$-plane or the $z$-plane.

Thus $z = \exp(s_1 T)$ where $s_1 = \sigma + j\omega$ and $\omega$ corresponds to frequencies on the $z$-plane, whereas $\Omega$ corresponds to frequencies on the $s$-plane.

## 4.3 The bilinear transformation

Kaiser[1] and Golden and Kaiser[2] suggest that the relationship between the complex variables $s$ and $s_1$, be of the form

$$s = \frac{2}{T} \tanh\left(\frac{s_1 T}{2}\right) \tag{4.1}$$

where $T$ is the sampling period.

Equation (4.1) states that the complex variable $s$ of the given continuous filter transfer function can be replaced by the hyperbolic tangent function of $s_1$.

Since the hyperbolic tangent function is a monotonic function it follows that all the values of $s$ are preserved in their correct order but, because of the periodic nature of the relationship, the $s$-plane is mapped onto the $s_1$-plane in a series of parallel strips as shown in Figure 4.1(a) and Figure 4.1(b). The doubly shaded area in Figure 4.1(b) situated around the origin is called the 'baseband region'. (In particular this term corresponds to frequencies on the imaginary axis within this region.)

From equation (4.1) we have,

$$\Omega = \frac{2}{T} \tan\left(\frac{\omega T}{2}\right) \tag{4.2}$$

so that the entire $\Omega$-axis is mapped monotonically in the regions $(2r - 1)\pi/T < \omega < (2r + 1)\pi/T$ for all integer values of $r$. Furthermore equation (4.1) can be expressed in terms of the complex variable $z^{-1}$ in the form

$$s = \frac{2}{T} \cdot \frac{1 - z^{-1}}{1 + z^{-1}} \tag{4.3}$$

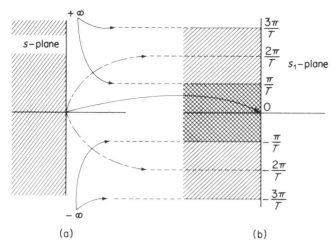

**Figure 4.1** The mapping of the bilinear transformation

so that when $s$ is replaced in a given real rational transfer function, by equation (4.3) we obtain a real rational function in $z^{-1}$. The amplitude characteristic corresponding to the continuous filter is reproduced in all the regions of $(2r - 1)\pi/T < \omega < (2r + 1)\pi/T$. Considering the baseband $-\pi/T < \omega < \pi/T$ we see that if the given continuous filter transfer function is a lowpass filter of cutoff frequency $\Omega_c$ then the resulting function of $z^{-1}$ corresponding to a digital filter of cutoff frequency $\omega_c$ is such that

$$\Omega_c = \frac{2}{T} \tan\left(\frac{\omega_c T}{2}\right). \tag{4.4}$$

This has the effect of changing the cut-off frequency of the continuous filter. It is usually referred to as a warping of the frequency scale and the relationship between the two cutoff frequencies of equation (4.4) is depicted in Figure 4.2.

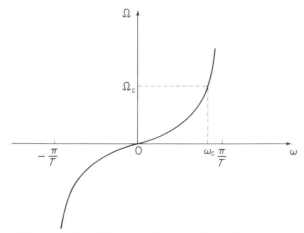

**Figure 4.2** The bilinear transformation for real frequencies

As a consequence of this warping one must use a continuous filter transfer function which has a particular frequency $\Omega_c$ to obtain a digital filter transfer function of the desired cutoff frequency $\omega_c$. In practice, however, we have tables of normalized continuous-filter transfer functions, i.e. $\Omega_c = 1 \text{ rad/s}$. Kaiser[1] and Kaiser and Golden[2] overcome this difficulty by using a lowpass to lowpass transformation (i.e. replacing the normalized variable $s$ by $s/\Omega_c$) before applying the transformation of equation (4.3).

The following two remarks can be made concerning the bilinear transformation of equation (4.3).

(i) The presence of the real constant multiplier $2/T$ is not necessary as it was pointed out by Rader and Gold[3], and,

(ii) The double transformation process (i.e. the normalized lowpass continuous to lowpass continuous of cutoff frequency $\Omega_c$ and then the bilinear transformation) involves a duplication of effort.

Rader and Gold[3] realized that the constant $2/T$ is superfluous and thus they used the transformation in the form

$$s = \frac{1 - z^{-1}}{1 + z^{-1}} \tag{4.5}$$

for which the second remark above is again applicable. Since any constant multiplier does not change the form of the transformation we can use the form

$$s = k\frac{1 - z^{-1}}{1 + z^{-1}} \tag{4.6}$$

where $k$ is a real constant and its value is such that,

$$\Omega_c = k \tan\left(\frac{\omega_c T}{T}\right) \tag{4.7}$$

where $\Omega_c$ is the cutoff frequency of the continuous filter, $\omega_c$ is the desired cutoff frequency of the digital filter. In particular if a table of normalized lowpass continuous filters is used then $\Omega_c = 1$ and

$$k = \cot\left(\frac{\omega_c T}{2}\right). \tag{4.8}$$

Hence the transformation of equation (4.6) becomes

$$s = \Omega_c \cdot \cot\left(\frac{\omega_c T}{2}\right) \cdot \frac{1 - z^{-1}}{1 + z^{-1}} \tag{4.9}$$

and for

$$\Omega_c = 1$$

$$s = \cot\left(\frac{\omega_c T}{2}\right) \cdot \frac{1 - z^{-1}}{1 + z^{-1}}. \tag{4.10}$$

Thus by incorporating a general constant $k$ in the transformation we have removed the constraint of having to transform twice.

The transformation of equation (4.9) therefore will give a lowpass digital filter of the desired cutoff frequency from a lowpass continuous filter.

If a highpass, bandpass or band-elimination digital filter is required then Kaiser[1] and others[2,3,4] suggest that the well known transformations of continuous filters can be applied to the normalized lowpass continuous filter before the bilinear transformation is used. This process will involve a superfluous amount of computation and as Rader and Gold[3] suggest for the case of bandpass digital filters, a new transformation can be used to obtain the digital filter directly from the lowpass continuous transfer function.

Let us consider the highpass case which can be obtained directly from the transformation already given in equation (4.5). If in a given lowpass continuous transfer function we were to replace $s$ by $1/s$ then we would obtain a highpass continuous transfer function. In view of equation (4.5) this means that if $s$ of the continuous lowpass filter is replaced by $(1 + z^{-1})/(1 - z^{-1})$ then we shall obtain a highpass digital filter.

In general, therefore, we have the lowpass continuous to highpass digital transformation been given in the form,

$$s = k\frac{1 + z^{-1}}{1 - z^{-1}}$$

where $k$ is a real positive constant whose value is such that the lowpass continuous filter cutoff frequency ($\Omega_c$) and the highpass digital filter cutoff frequency ($\omega_c$) are related through the equation

$$k = \Omega_c . \tan\left(\frac{\omega_c T}{2}\right).$$

Hence the transformation becomes

$$s = \Omega_c . \tan\left(\frac{\omega_c T}{2}\right) . \frac{1 + z^{-1}}{1 - z^{-1}}.$$

If the lowpass continuous filter is normalized, i.e. if $\Omega_c = 1$, then we have

$$s = \tan\left(\frac{\omega_c T}{2}\right) . \frac{1 + z^{-1}}{1 - z^{-1}}.$$

In either case above, the transformation is in such a form that it obviates the need to

(i) Prewarp (this process is included in the real constant $k$).

(ii) Transform the lowpass continuous into a highpass continuous filter transfer function.

The advantages of this transformation from the point of view of minimizing computation are, therefore, immediately apparent.

The transformation of Rader and Gold is a special case corresponding to one-passband in the baseband of the general multiband transformations given in the next section. The prewarping procedure suggested by Rader and Gold is dispensed with if a modified form of the transformation is used. The following section deals with the general transformations and with the specific forms for the bandpass and band elimination digital filters. The necessary design formulae are also given.

### 4.4 The general lowpass continuous to multiband filter transformations

Let

$$\alpha_i = \cos \omega_{0i} T$$

$$\beta_i = \cos \omega_{\infty i} T$$

and

$$0 < \omega_{00} < \omega_{\infty 0} < \omega_{01} < \omega_{\infty 1} \cdots \omega_{0(n-1)} < \frac{\pi}{T}$$

where $T$ is the sampling period. Then we have the following result.

The transformation

$$s = k \cdot \frac{1}{1 - z^{-2}} \cdot \frac{\displaystyle\prod_{i=0}^{n-1} (z^{-2} - 2\alpha_i z^{-1} + 1)}{\displaystyle\prod_{i=0}^{n-2} (z^{-2} - 2\beta_i z^{-1} + 1)}, \qquad n \geqslant 2 \qquad (4.11)$$

transforms a lowpass continuous filter transfer function into a multiband pass digital filter transfer function where the number of passbands is $(n - 1)$ and $k$ is a real positive constant.

Thus for one passband in the baseband we obtain the transformation,

$$s = k \cdot \frac{z^{-2} - 2\alpha z^{-1} + 1}{1 - z^{-2}} \qquad (4.12)$$

where

$$\alpha = \alpha_0 = \cos \omega_{00} T.$$

The transformation of equation (4.12) has been given by Rader and Gold[3] with the real constant $k$ omitted and hence its application necessitated a prewarping procedure. The inclusion of $k$, however, obviates this need, as is shown below.

Let the lowpass continuous filter have a cutoff frequency $\Omega_c$ and let the desired upper and lower cutoff frequencies of the bandpass digital filter be $\omega_2$ and $\omega_1$ respectively.

Then, when $z^{-1} = \exp(-j\omega T)$, i.e. when $z^{-1}$ varies on the unit circle[5,6] we have

$$\Omega = k \cdot \frac{\alpha - \cos \omega T}{\sin \omega T} \tag{4.13}$$

so that the point $\Omega = 0$ corresponds to the point

$$\omega = \frac{1}{T} \cos^{-1} \alpha = \omega_{00},$$

i.e. $\omega_{00}$ is the centre frequency of the band.

Furthermore,

$$-\Omega_c = k \frac{\alpha - \cos \omega_1 T}{\sin \omega_1 T}$$

and $\tag{4.14}$

$$\Omega_c = k \frac{\alpha - \cos \omega_2 T}{\sin \omega_2 T}$$

which, after simplification, give

$$\alpha = \frac{\cos \left( \dfrac{\omega_2 + \omega_1}{2} \right) T}{\cos \left( \dfrac{\omega_2 - \omega_1}{2} \right) T} \tag{4.15}$$

and

$$k = \Omega_c \cdot \cot \left( \frac{\omega_2 - \omega_1}{2} \right) T. \tag{4.16}$$

Thus, having specified the upper and lower cutoff frequencies of the bandpass digital filter, one can choose the cutoff frequency of the lowpass continuous filter and the constant $k$ to satisfy equation (4.16). There is no need, therefore, to prewarp the lowpass continuous filter transfer function.

Now referring back to equation (4.11) we can make the following observation. If, instead of replacing $s$ by the function of equation (4.11), we were to replace $1/s$, then we would have a multiband elimination transformation. We can formulate this, therefore, as follows.

The transformation

$$s = k(1 - z^{-2}) \cdot \frac{\displaystyle\prod_{i=0}^{n-2} (z^{-2} - 2\alpha_i z^{-1} + 1)}{\displaystyle\prod_{i=0}^{n-1} (z^{-2} - 2\beta_i z^{-1} + 1)}, \qquad n \geqslant 2 \tag{4.17}$$

transforms a lowpass continuous filter transfer function into a multiband elimination digital filter transfer function where the number of elimination bands is $(n - 1)$, $k$ is a real positive constant, and

$$\alpha_i = \cos \omega_{0i} T$$

$$\beta_i = \cos \omega_{\infty i} T$$

$$0 < \omega_{\infty 0} < \omega_{00} < \omega_{\infty 1} < \cdots \omega_{\infty(n-1)} < \frac{\pi}{T}.$$

For the case of one elimination band we have from the above expression,

$$s = k \cdot \frac{1 - z^{-2}}{z^{-2} - 2\alpha z^{-1} + 1} \tag{4.18}$$

where

$$\alpha = \alpha_0 = \cos \omega_{\infty 0} T.$$

When $z^{-1}$ is on the unit circle then we have,

$$\Omega = k \frac{\sin \omega T}{\cos \omega T - \alpha}. \tag{4.19}$$

The centre of the attenuation band is situated at the frequency $1/T \cos^{-1} \alpha$, i.e. it is given by

$$\omega = \omega_{\infty 0} = \frac{1}{T} \cos^{-1} \alpha.$$

Let the upper and lower cutoff frequencies of the band-elimination digital filter be $\omega_2$ and $\omega_1$, and the cutoff frequency of the lowpass continuous filter be $\Omega_c$. Then the following relationships must hold.

$$\Omega_c = k \frac{\sin \omega_1 T}{\cos \omega_1 T - \alpha}$$

and $$\tag{4.20}$$

$$-\Omega_c = k \cdot \frac{\sin \omega_2 T}{\cos \omega_2 T - \alpha}.$$

From the above relationships we obtain,

$$\alpha = \frac{\cos \left( \frac{\omega_2 + \omega_1}{2} \right) T}{\cos \left( \frac{\omega_2 - \omega_1}{2} \right) T}$$

which is identical to equation (4.15), and also

$$k = \Omega_c \cdot \tan \left( \frac{\omega_2 - \omega_1}{2} \right) T. \tag{4.21}$$

The remarks made for the bandpass case are equally applicable to the above case also. Thus no prewarping is necessary since there are two degrees of freedom in equation (4.21).

## 4.5 Remarks

In Section 4.3 we have reviewed and modified the bilinear transformation method of synthetizing digital filters, whereas in Section 4.5 we have given two multiband transformations. One corresponds to the multibandpass case from which a more general form of the Rader and Gold[3] transformation was obtained obviating the need to prewarp the continuous filter transfer function. The other general transformation corresponds to the multi-band elimination case, from which the 'one-elimination band' transformation was derived.

A special form of the transformation of equation (4.12) corresponds to the case when $\alpha = 0$ and $k = 1$, i.e. when

$$\omega_2 + \omega_1 = \frac{\pi}{T}$$

and

$$\omega_2 - \omega_1 = \frac{2}{T} \tan^{-1} \Omega_c$$

Then

$$s = \frac{1 + z^{-2}}{1 - z^{-2}}$$

and

$$\omega_2 = \tfrac{1}{2} \left[ \frac{\pi}{T} + \frac{2}{T} \tan^{-1} \Omega_c \right]$$

$$\omega_1 = \tfrac{1}{2} \left[ \frac{\pi}{T} - \frac{2}{T} \tan^{-1} \Omega_c \right]$$

In this case we have an exactly arithmetically symmetrical bandpass digital filter.

Another special form corresponds to the case when $k = 1$ which gives

$$\omega_2 - \omega_1 = \frac{2}{T} \tan^{-1} \Omega_c.$$

This form corresponds to the Rader and Gold[3] transformation and as can be seen from the above equation is rather restrictive in either the choice of the upper and lower cutoff frequencies or the cutoff frequency of the continuous lowpass digital filter. It implies that by specifying the cutoff frequencies of the band pass filter we must use a lowpass continuous filter whose cutoff frequency satisfies the above condition and since our aim is to use tabulated filters (i.e. $\Omega_c = 1$) we must apply a lowpass continuous to lowpass continuous transformation before applying the Rader and Gold form.

Consider the case when $\alpha = 0$ in both equation (4.12) and equation (4.18). Then equation (4.12) becomes

$$s = k\left(\frac{1 + z^{-2}}{1 - z^{-2}}\right). \tag{4.22}$$

Comparing equation (4.22) with equation (4.6) it is seen that if we replace $z^{-1}$ in equation (4.6) by $(-z^{-2})$ then we obtain equation (4.22). This result is very important since the special case of the arithmetically symmetrical bandpass digital filter is obtained as it was given elsewhere[6,7], similarly, with the band elimination case we obtain from equation (4.18)

$$s = k\left(\frac{1 - z^{-2}}{1 + z^{-2}}\right) \tag{4.23}$$

which is equivalent to replacing $z^{-1}$ by $z^{-2}$. (It is understood that in the above cases the corresponding constants $k$ are identical.)

Thus the two special forms of the spectral transformations on the $z^{-1}$ plane can be derived from the above functions.

Table 4.1   $s$-plane to $z^{-1}$-plane transformations

Given a lowpass continuous transfer function of cut-off frequency $\Omega_c$

| Required digital filter | Cut-off frequency | Replace $s$ by: | Where: |
|---|---|---|---|
| Lowpass | $\omega_c$ | $k \cdot \dfrac{1 - z^{-1}}{1 + z^{-1}}$ | $k = \Omega_c \cot \dfrac{\omega_c T}{2}$ |
| Highpass | $\omega_c$ | $k \cdot \dfrac{1 + z^{-1}}{1 - z^{-1}}$ | $k = \Omega_c \tan \dfrac{\omega_c T}{2}$ |
| Bandpass | $\omega_1, \omega_2$ centre frequency $\omega_0$   $\alpha = \cos \omega_0 T$ | $k \cdot \dfrac{z^{-2} - 2\alpha z^{-1} + 1}{1 - z - 2}$ | $\alpha = \dfrac{\cos\left(\dfrac{\omega_2 + \omega_1}{2}\right) T}{\cos\left(\dfrac{\omega_2 - \omega_1}{2}\right) T}$   $k = \Omega_c \cot\left(\dfrac{\omega_2 - \omega_1}{2}\right) T$ |
| Bandstop | $\omega_1, \omega_2$ centre frequency $\omega_0$   $\alpha = \cos \omega_0 T$ | $k \cdot \dfrac{1 - z^{-2}}{z^{-2} - 2\alpha z^{-1} + 1}$ | $\alpha = $ As above   $k = \Omega_c \cdot \tan\left(\dfrac{\omega_2 - \omega_1}{2}\right) T$ |

## 4.6   Summary of transformations

The transformations are given in Table 4.1 along with the necessary design formulae.

These transformations, since they are algebraic in nature, can be applied to any form of the continuous filter transfer function whether it is in partial fraction form or as a ratio of polynomials in $s$.

## 4.7   Example

A lowpass digital filter is required to be synthetized having the following specifications.

$$\begin{aligned}
\text{Cutoff frequency} &= 6\,\text{kHz} \\
\text{Transition frequency} &= 8.8\,\text{kHz} \\
\text{Maximum passband attenuation} &= 1\,\text{dB} \\
\text{Minimum stopband attenuation} &= 30\,\text{dB}
\end{aligned}$$

The sampling frequency is to be 32 kHz.

The tolerance scheme for the digital filter is shown in Figure 4.3(a). To find the equivalent specifications for the continuous filter we apply equation (4.10) as follows,

(i) The transition frequency $\Omega_1$ will be given by,

$$\Omega_1 = \cot\left(\frac{2\pi \times 6 \times 10^3 \times 1}{32 \times 10^3 \times 2}\right) \times \tan\left(\frac{2\pi \times 8.8 \times 10^3 \times 1}{32 \times 10^3 \times 2}\right)$$

$$\simeq 1.75\,\text{rad/s}.$$

Therefore the tolerance scheme for the continuous filter is as shown in Figure 4.3(b). Note that the frequency axes between the two figures are not related linearly. With reference to the scheme of Figure 4.3(b) and with the aid of tabulated filter[8,9] transfer functions, we obtain the following function that satisfies the requirements

$$H(s) = \frac{(1 + as)}{(1 + bs)(1 + cs + ds^2)}$$

where

$$a = 0.25517931$$

$$b = 1.793438$$

$$c = 0.4066687$$

$$d = 0.9845149$$

58

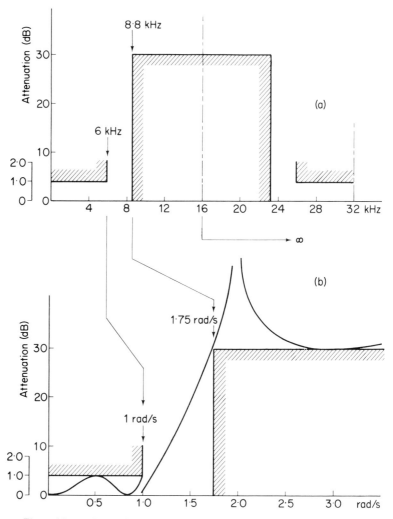

**Figure 4.3** (a) Desired attenuation scheme; (b) realizable filter characteristic

The attenuation characteristic is indicated on Figure 4.3(b).

Now the required transformation according to equation (4.10) will be given by,

$$s = \cot\left(\frac{2\pi \times 6 \times 10^3}{32 \times 10^3 \times 2}\right)\frac{1 - z^{-1}}{1 + z^{-1}}$$

or

$$s = 1.4966058 \frac{1 - z^{-1}}{1 + z^{-1}}.$$

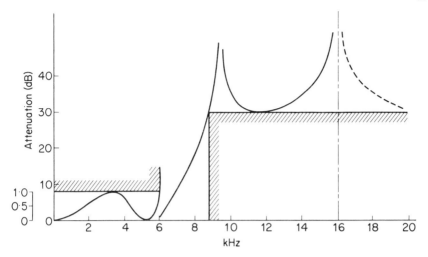

**Figure 4.4**   Digital filter characteristic

Thus on replacing $s$ in $H(s)$ by the above expression we obtain the required result

$$G(z) = H(s)|_{s = 1 \cdot 4966058(1 - z^{-1})/(1 + z^{-1})}$$

or

$$G(z) = \frac{(1 + z^{-1})(1 \cdot 571558 + 0 \cdot 856884z^{-1} + 1 \cdot 571558z^{-2})}{(3 \cdot 684070 - 1 \cdot 684070z^{-1})(3 \cdot 813767 - 2 \cdot 410288z^{-1} + 2 \cdot 596521z^{-2})}.$$

The attenuation characteristic of $G(z)$ is shown in Figure 4.4 where the tolerance scheme is also indicated for comparison.

### References

1. Kuo, F. F., and Kaiser, J. F., *Systems Analysis by Digital Computer*, Wiley, 1966.
2. Golden, R. M., and Kaiser, J. F., 'Design of wideband sampled data filters', *B.S.T.J.*, Vol. **43** Part 2, 1533–1546 (1964).
3. Rader, C. M., and Gold, B., *Digital Methods for Sampled-Data Filters*, Proceedings First Allerton Conference on Circuit and System Theory, 1963, pp. 221–236.
4. Gold, B., and Rader, C. M., *Digital Processing of Signals*, McGraw-Hill, 1969.
5. Constantinides, A. G., 'Frequency transformations for digital filters', *Electron Letters*, **3** No. 11, 487–489 (1967).
6. Constantinides, A. G. 'Frequency transformations for digital filters', *Electron Letters* **4** No. 7, 115–116 (1968).
7. Constantinides, A. G., *Digital Filters in Frequency Division Multiplex Telephone Systems*, Research Report, The City University, London, June 1969.
8. Saal, R., *Der Entwurf von Filtern mit Hilfe des Kataloges normierter Tiefpasse*, Telefunken AG, 1967.
9. Christian, E., and Eisenmann, E., *Filter Design Tables and Graphs*, John Wiley, New York, 1966.

*Chapter 5*

# Direct Synthesis of Digital Filters

*A. G. Constantinides*

## 5.1 Introduction

The problem of direct synthesis of digital filters is divided into two problems in a manner similar to that of continuous filters.

(a) Direct synthesis of lowpass digital filters and,

(b) direct synthesis of highpass, bandpass and bandstop digital filters from lowpass filter transfer functions.

This division simplifies the problem considerably, because if highpass, bandpass and band-elimination transfer functions are obtainable from lowpass transfer functions then it is only necessary to synthesize lowpass transfer functions.

The ideal amplitude characteristic of a lowpass digital filter has the brick-wall shape as given in Figure 5.1. The square of the amplitude characteristic of a realizable digital filter, however, will be shown to be a real rational function in $\tan^2(\omega T/2)$.

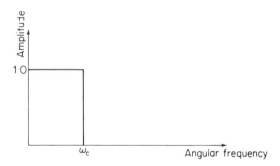

**Figure 5.1** Ideal lowpass characteristic

Thus the problem reduces to that of finding a particular function which approximates to the ideal amplitude characteristic in a prescribed sense. Approaching the problem in this way is direct in that only conditions pertinent to digital filters are used, starting from the definition of the ideal amplitude characteristic.

Depending on the sense of approximating this ideal characteristic, the resulting pulse transfer functions are classified as follows:

   (i) Maximally flat or Butterworth approximation
  (ii) Chebyshev approximation[2]
 (iii) Elliptic type[3]
 (iv) Generalized polynomial approximation[5]
  (v) General parameter type.[6]

In this section lowpass digital filters belonging to class (i) are studied and synthesis procedures put forward.

The first step to the direct synthesis of digital filters is to show that their pulse transfer functions can be constructed from given amplitude characteristics.

The general form of the square of amplitude characteristics will be as follows.

*Amplitude characteristic:* The transfer function of a digital filter was shown in Chapter 3 (equation 3.5) to be of the form,

$$G(z) = \frac{\sum_r a_r z^{-r}}{\sum_r b_r z^{-r}}.$$

(5.1)

When $z^{-1} = \cos \omega T - j \sin \omega T$ the above equation becomes

$$G(\omega T) = \frac{\sum_r a_r \cos r\omega T - j \sum_r a_r \sin r\omega T}{\sum_r b_r \cos r\omega T - j \sum_r b_r \sin r\omega T}.$$

(5.2)

Now consider the expression,

$$y_n = \cos n\theta \text{ where } n \text{ is an integer.}$$

Then

$$y_{n-1} = \cos(n-1)\theta = \cos n\theta \cos n\theta + \sin n\theta \sin \theta$$

and

$$y_{n+1} = \cos(n+1)\theta = \cos n\theta \cos n\theta - \sin n\theta \sin \theta$$

so that $y_{n+1} + y_{n-1} = 2 \cos \theta y_n$ and hence,

$$y_{n+1} = 2 \cos \theta \, y_n - y_{n-1}.$$

(5.3)

But $y_0 = 1$ and $y_1 = \cos \theta$ hence from the recurrence relationship of equation (5.3) it follows that $y_n$ is a real polynomial in $\cos \theta$.

Now consider

$$u_n = \sin n\theta.$$

Then

$$u_{n+1} = \sin(n+1)\theta = \sin n\theta \cos \theta + \cos n\theta \sin \theta$$

$$u_{n-1} = \sin(n-1)\theta = \sin n\theta \cos \theta - \cos n\theta \sin \theta.$$

Therefore,

$$u_{n+1} - u_{n-1} = 2 \sin \theta \cos n\theta$$

$$= 2 \sin \theta \, y_n$$

and since $y_n$ is a polynomial in $\cos\theta$ it follows that $u_n$ must be of the form $\sin\theta \times$ (polynomial in $\cos\theta$).

With these results in mind let us return to equation (5.2). It can be seen that the real parts of the numerator and denominator are summations of cosines of the multiple angle $\omega T$ and hence they can be expressed as polynomial in $\cos\omega T$. The imaginary parts however are summations of sines of the multiple angle $\omega T$ and hence they will have a common factor $\sin\omega T$ which is multiplied by polynomials in $\cos\omega T$. Hence equation (5.2) can be written in the form,

$$G(\omega T) = \frac{\sum_r c_r \cos^r \omega T - j \sin\omega T \sum_r d_r \cos^r \omega T}{\sum_r e_r \cos^r \omega T - j \sin\omega T \sum_r f_r \cos^r \omega T}$$

and hence the amplitude characteristic has its square in the form,

$$|G(\omega T)|^2 = \frac{\left(\sum_r c_r \cos^r \omega T\right)^2 + (1 - \cos^2 \omega T)\left(\sum_r d_r \cos^r \omega T\right)^2}{\left(\sum_r e_r \cos^r \omega T\right)^2 + (1 - \cos^2 \omega T)\left(\sum_r f_r \cos^r \omega T\right)^2}$$

$$= \frac{\sum_r g_r \cos^r \omega T}{\sum_r h_r \cos^r \omega T}$$

or by letting

$$\cos\omega T = \frac{1 - \tan^2 (\omega T/2)}{1 + \tan^2 (\omega T/2)}$$

we obtain,

$$|G(\omega T)|^2 = \frac{\sum_r \alpha_r \tan^{2r} (\omega T/2)}{\sum_r \beta_r \tan^{2r} (\omega T/2)} \tag{5.4}$$

where $\alpha_r$ and $\beta_r$ are real constant coefficients. *Thus, the square of the amplitude characteristic of a digital filter can be expressed as a real rational function in* $\tan^2 (\omega T/2)$.

Equation (5.4) can always be expressed in the form,

$$|G(\omega T)|^2 = \frac{1}{1 + \varepsilon^2 P^2(\omega)} \tag{5.5}$$

where $\varepsilon$ is a real constant and $P^2(\omega)$ is a real rational function in $\tan^2 (\omega T/2)$. This form is eminently suitable for filter characteristics.

We shall be concerned with lowpass characteristics only under the assumption that highpass bandpass and band-elimination characteristics can be derived from lowpass prototype.

Thus $P^2(\omega)$ is necessarily a $(2m)$th order in $z^{-1}$ in both the numerator and denominator and therefore $G(z^{-1})$ has its numerator and denominator in $z^{-1}$ of equal degree which is the degree of the pulse transfer function.

## 5.2 Polynomial lowpass digital filters

In general, the squared amplitude characteristic is expressed in the form,

$$|G|^2 = \frac{1}{1 + \varepsilon^2 P^2(\omega)} \tag{5.6}$$

as pointed out above.

By the term 'polynomial filters', it is meant that the function $P^2(\omega)$ of equation (5.6) is expressed in the general form,

$$P^2(\omega) = \sum_{r=0}^{n} a_r \tan^{2r}\left(\frac{\omega T}{2}\right) \tag{5.7}$$

where $\tan(\omega T/2)$ is the lowpass frequency variable and all $a_r$ are real constant coefficients.

Since equation (5.6) is to approximate to the ideal amplitude characteristic, with $P^2(\omega)$ given in the form of equation (5.7), then some specific attributes of the sense of approximation can be imposed on (5.7) to yield the prescribed form $|G|^2$. Before proceeding to the evaluation of the various forms of $P^2(\omega)$, let us examine its general form, as given by equation (5.7) and formulate its basic properties.

It is seen that at the frequencies $(2n + 1)(\Omega_s/2)$, where $\Omega_s$ is the angular sampling frequency, (and hence at the Nyquist limits also), we have,

$$\tan\left[\frac{T}{2} \cdot (2n + 1) \cdot \frac{\Omega_s}{2}\right] = \pm\infty \tag{5.8}$$

hence $P^2(\omega)$ is infinite at these frequencies and therefore the amplitude characteristic is zero.

Further, when the function $\tan(\omega T/2)$ is large, then equation (5.7) behaves as,

$$P^2(\omega) = 0(\tan^{2n}(\omega T/2)) \tag{5.9}$$

and since $\tan(\omega T/2)$ is a monotonic function of $\omega$, $P^2(\omega)$ and therefore $|G|^2$, will vary in a monotonic fashion with $\omega$. That is, $P^2(\omega)$ will increase monotonically to infinity and hence $|G|^2$ will decrease monotonically to zero at $(2n + 1)\Omega_s/2$.

An obvious remark is that $|G|^2$ of equation (5.6) is periodic with $\omega$, having a period of $2\pi/T = \Omega_s$. In view of this we can concentrate our attention within one period, say the baseband $(-\Omega_s/2, \Omega_s/2)$, for which the following point is noted.

*The squared amplitude characteristic $|G|^2$ of equation (5.6), with $P^2(\omega)$ given by equation (5.7), decreases monotonically above a certain frequency $\omega_c$, to zero at $\Omega_s/2$.*

It is interesting to compare the above property with that of a continuous lowpass filter, where $P^2(\omega)$ is a real polynomial in $\omega^2$, and which above a certain frequency becomes a monotonically increasing function, becoming infinite at infinite frequency. Thus for an $n$th order continuous filter of this type there is an $n$th order zero of the transfer function at infinity, whereas for an $n$th order digital filter function of the type (5.7) there is an $n$th order zero of the transfer function at the Nyquist frequency in continuous and digital filters, respectively, which serve the same purpose.

Dividing the baseband into a passband and a stopband it is seen that the choice of $P^2(\omega)$ as given by equation (5.7) makes the amplitude characteristics of lowpass digital filters monotonic in the stopband. The passband variation, however, must be determined from its prescribed nature, i.e. Monotonic Butterworth-type, equiripple Chebyshev-type, etc.

As a final property, it is observed that as $n \to \infty$ in equation (5.7), then equation (5.6) approaches the ideal shape.

### 5.3 Synthesis of monotonic Butterworth-type lowpass digital filters

The simplest non-trivial form of $P^2(\omega)$ of equation (5.7) is,

$$P^2(\omega) = a \cdot \tan^{2n}\left(\frac{\omega T}{2}\right) \tag{5.10}$$

where $a$ is a real constant coefficient.

This function is monotonic in the passband and stopband.

The amplitude characteristic of a lowpass filter is then given by,

$$|G| = 1 \bigg/ \sqrt{1 + a \tan^{2n}\left(\frac{\omega T}{2}\right)}$$

and if it is to fall by 3 dB at the cutoff frequency $\omega_c$, then the constant $a$ becomes

$$a = \left[\tan^{2n}\frac{\omega_c T}{2}\right]^{-1}$$

and therefore the square of the above amplitude characteristic becomes,

$$|G(\omega)|^2 = 1 \bigg/ \left\{1 + \left[\frac{\tan(\omega T/2)}{\tan(\omega_c T/2)}\right]^{2n}\right\}. \tag{5.11}$$

Thus for equation (5.11) we have,

$$\text{(i)} \quad |G(0)| = 1$$

$$\text{(ii)} \quad |G(\omega_c)| = 1/\sqrt{2}$$

$$\text{(iii)} \quad G\left(\frac{\pi}{T}\right) = 0.$$

The amplitude characteristic of (5.11) decreases, therefore, monotonically from unity to zero for the baseband frequencies in a sense similar to the Butterworth-type lowpass continuous filter.

To synthesize the lowpass filter having this amplitude characteristic, it is necessary to determine the location of the poles on the $z^{-1}$-plane.

To this end let

$$p = \frac{1 - z^{-1}}{1 + z^{-1}} \tag{5.12}$$

where $p = u + jv$ and $z^{-1} = e^{-j\omega T}$ on $C:|z^{-1}| = 1$, so that when $z^{-1}$ lies on the unit circle, the auxiliary complex variable $p$ becomes

$$p = -j \tan\left(\frac{\omega T}{2}\right)$$

or

$$u = 0 \quad \text{and} \quad v = -\tan\left(\frac{\omega T}{2}\right). \tag{5.13}$$

On the $p$-plane, the poles of equation (5.11) will be located at the roots of the equation,

$$\left(\frac{jp}{v_c}\right)^{2n} + 1 = 0$$

where $v_c = \tan(\omega_c T/2)$.

$\omega_c$ is the required cutoff frequency, or

$$(-1)^n \cdot \left(\frac{p}{v_c}\right)^{2n} + 1 = 0.$$

Thus for even $n$,

$$p_r = v_c e^{j((2r+1)/2n)\pi} \qquad r = 0, 1, 2, \ldots, (2n - 1)$$

and for odd $n$

$$p_r = v_c e^{j(r\pi/n)} \qquad r = 0, 1, 2, \ldots, (n - 1).$$

Hence, for $n$ even,

$$u_r = v_c \cos\left(\frac{2r + 1}{2n}\right)\pi$$

$$r = 0, 1, 2, \ldots, (2n - 1) \tag{5.14}$$

$$v_r = v_c \sin\left(\frac{2r + 1}{2n}\right)\pi$$

and for $n$ odd replace $(2r + 1)/2n$ in the above equations by $r/n$.

The parametric equations of (5.14) describe a circle at the origin of radius $v_c$ on the $p$-plane. This circle on the $z^{-1}$-plane through the transformation (5.12) becomes the following curve:

Let

$$z^{-1} = x + jy$$

and

$$p = u + jv$$

so that

$$z^{-1} = x + jy = \frac{1 - (u + jv)}{1 + (u + jv)}$$

hence

$$x = \frac{1 - (u^2 + v^2)}{(1 + u)^2 + v^2}$$

$$y = \frac{-2v}{(1 + u)^2 + v^2}.$$

But

$$u^2 + v^2 = v_c^2$$

hence

$$x = \frac{1 - v_c^2}{1 + 2u + v_c^2}$$

$$y = \frac{-2v}{1 + 2u + v_c^2} \qquad (5.15)$$

from which we obtain the relationship between $x$ and $y$ as

$$y^2 + \left[ x - \frac{1 + v_c^2}{1 - v_c^2} \right]^2 = \frac{4v_c^2}{(1 - v_c^2)^2}.$$

The above equation describes a circle of radius

$$\rho = 2v_c/(1 - v_c^2)$$

and centre C

$$\left[ \frac{1 + v_c^2}{1 - v_c^2}, 0 \right].$$

Since

$$v_c = -\tan\left( \frac{\omega_c T}{2} \right).$$

The radius and centre of the circle become,

$$\rho = \tan(\omega_c T), \qquad C(\sec \omega_c T, 0).$$

From equations (5.14) and (5.15), the pole positions are then given by,

$$x_r = \frac{1 - \tan^2\left(\frac{\omega_c T}{2}\right)}{1 - 2\tan\left(\frac{\omega_c T}{2}\right).\cos\left(\frac{2r+1}{2n}\right)\pi + \tan^2\left(\frac{\omega_c T}{2}\right)}$$

$$y_r = \frac{2\tan\left(\frac{\omega_c T}{2}\right)\sin\left(\frac{2r+1}{2n}\right)\pi}{1 - 2\tan\left(\frac{\omega_c T}{2}\right).\cos\left(\frac{2r+1}{2n}\right)\pi + \tan^2\left(\frac{\omega_c T}{2}\right)}$$

$$r = 0, 1, 2, \ldots, (2n-1)$$

for even $n$, whereas for odd $n$, the above equations hold with $((2r+1)/2n)$ replaced by $(r/n)$ and the range of $r$ becomes, $r = 0, 1, 2, 3, \ldots (n-1)$.

The synthesis procedure will be as follows:

1. From the given specifications evaluate the order of the filter $n$.
2. Determine the pole positions on the $z^{-1}$-plane and choose those that lie outside the unit circle.
3. It is known from equation (5.8) that there exists an $n$th order zero at $z^{-1} = -1$. From this, and from the poles found in step 2, construct the required transfer function.

Consider the following example.

**Example.** A lowpass digital filter is required to have a maximally flat amplitude characteristic of cutoff frequency 4·5 kHz. The transition ratio is required to be 0·9 and the attenuation at the transition frequency better than 60 dB. The sampling frequency is 18 kHz.

Evaluation of the order $n$.

From the transition ratio and cutoff frequency, the transition frequency is calculated,

$$f_1 = 4\cdot5/0\cdot9$$

i.e.

$$f_1 = 5\,\text{kHz}$$

and hence $\omega_1 = 2\pi f_1 = 10\pi\,\text{krad/s}$. Now,

$$\tan\left(\frac{\omega_c T}{2}\right) = \tan\left(\frac{4\cdot5\pi}{18}\right)$$

$$= 1$$

and

$$\tan\left(\frac{\omega_1 T}{2}\right) = \tan\left(\frac{5\pi}{18}\right) = 1\cdot18.$$

Hence at the transition frequency we have,

$$10 \log (1 + 1 \cdot 18^{2n}) = 60.$$

Therefore

$$n = \frac{6}{0 \cdot 79 \times 2} = 3 \cdot 8$$

and taking the nearest integer, $n = 4$.

Since $\tan (\omega_c T/2) = 1$, it follows that the poles on the $p$-plane lie on a unit circle, and in view of equation (5.14), they will be given by,

$$\left. \begin{aligned} u_r &= \cos \left( \frac{2r + 1}{8} \right) \pi \\ v_r &= \sin \left( \frac{2r + 1}{8} \right) \pi \end{aligned} \right\} . r = 0, 1, 2, 3, 4, 5, 6, 7$$

as given in the table below:

| $u_r$ | $v_r$ |
|---|---|
| $\pm 0 \cdot 92388$ | $\pm 0 \cdot 38268$ |
| $\pm 0 \cdot 38268$ | $\pm 0 \cdot 92388$ |

On the $z^{-1}$-plane these poles will lie on the imaginary axis and they will be at,

| $x_r$ | $y_r$ |
|---|---|
| 0 | $\pm 0 \cdot 19891$ |
| 0 | $\pm 5 \cdot 02732$ |
| 0 | $\pm 0 \cdot 66818$ |
| 0 | $\pm 1 \cdot 49660$ |

Figure 5.2 shows the pole distribution on the $z^{-1}$-plane. It is immediately noticed that four poles lie inside the unit circle of the $z^{-1}$-plane and the other four poles lie outside the unit circle. Thus the poles yielding a stable pulse transfer function are,

$$0 \pm j5 \cdot 02732$$

$$0 \pm j1 \cdot 49660.$$

Each set of the above poles will give a quadratic with the coefficient of $z^{-1}$ zero. Therefore the quadratic in $z^{-1}$ corresponding to the $p$-plane pair of poles $(-\cdot 38268 \pm j \cdot 92388)$ is given by

$$Q_1 = 1 \cdot 23464 z^{-2} + 2 \cdot 76536$$

and for the $p$-plane pair of poles $(-\cdot 92388 \pm \cdot 38268)$

$$Q_2 = 0 \cdot 15224 z^{-2} + 3 \cdot 84776$$

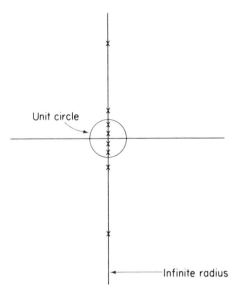

**Figure 5.2** Pole location for Butterworth filter
$n = 4, \omega_c T = \pi/2$

Since the filter is 4th order, there is a 4th order zero at $z^{-1} = -1$. Hence the required pulse transfer function is given by

$$G_1(z^{-1}) = K \cdot \frac{(1 + z^{-1})^4}{Q_1 \cdot Q_2}$$

where $K$ is the normalizing constant.
Thus, normalizing at $\omega = 0, (z^{-1} = 1)$ we have,

$$1 = K \cdot \frac{2^4}{4 \times 4}$$

i.e. $K = 1$.
The amplitude characteristic of $G(z^{-1})$ is shown in Figure 5.3.
Similarly Chebyshev-type and Elliptic (Cauer) type filters may be constructed[1,2,3].

## 5.4 Frequency transformations[4]

The problem of designing highpass, bandpass and bandstop filters from lowpass filters can be dealt with most efficiently through frequency transformations. We give in Table 5.1 the transformations which are necessary to transform a given lowpass digital filter to other forms. Some examples of the transformations are shown in Figure 5.4.

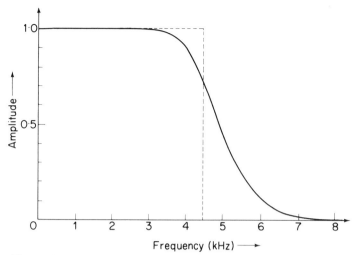

**Figure 5.3** Amplitude characteristic of Butterworth filter $n = 4$, $\omega_c T = \pi/2$

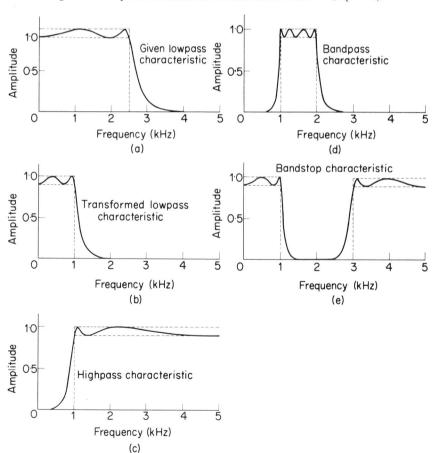

**Figure 5.4** Examples produced by frequency transformations

72

## References

1. Gold, B., and Rader, C. M., *Digital Processing of Signals*, McGraw-Hill, 1969.
2. Constantinides, A. G., 'Synthesis of Chebyshev digital filters', *Electron Letters*, **3** No. 3 (March 1967).
3. Constantinides, A. G., ' Elliptic digital filters', *Electron Letters*, **3** No. 6 (June 1967).
4. Constantinides, A. G., 'Spectral transformations for digital filters', *Proc. IEEE*, **117** No. 8 (August 1970).
5. Constantinides, A. G., 'Family of Equiripple Lowpass Digital Filters', *Electron Letters*, **6**, No. 11 (May 1970).
6. Deczky, A. G., 'Synthesis of Recursive Digital Filters using the Minimum p-Error Criterion', *IEEE Trans. Audio and Electroacoustics*, **AU-20**, No. 4 (October 1972).

Table 5.1   Frequency transformations from a lowpass-digital-filter prototype of cutoff frequency $\beta$

| Filter type | Transformation | Associated design formulae |
|---|---|---|
| Lowpass | $\dfrac{z^{-1} - \alpha}{1 - \alpha z^{-1}}$ | $\alpha = \dfrac{\sin\left(\dfrac{\beta - \omega_c}{2}\right)T}{\sin\left(\dfrac{\beta + \omega_c}{2}\right)T}$ |
| Highpass | $\dfrac{z^{-1} + \alpha}{1 + \alpha z^{-1}}$ | $\alpha = -\dfrac{\cos\left(\dfrac{\beta - \omega_c}{2}\right)T}{\cos\left(\dfrac{\beta + \omega_c}{2}\right)T}$ |
| Bandpass | $-\left(\dfrac{z^{-2} - \dfrac{2\alpha k}{k+1}z^{-1} + \dfrac{k-1}{k+1}}{\dfrac{k-1}{k+1}z^{-2} - \dfrac{2\alpha k}{k+1}z^{-1} + 1}\right)$ | $\alpha = \cos\omega_0 T = \dfrac{\cos\left(\dfrac{\omega_2 + \omega_1}{2}\right)T}{\cos\left(\dfrac{\omega_2 - \omega_1}{2}\right)T}$  $k = \cot\left(\dfrac{\omega_2 - \omega_1}{2}\right)T_0 \tan\dfrac{\beta T}{2}$ |
| Bandstop | $\left(\dfrac{z^{-2} - \dfrac{2\alpha}{1+k}z^{-1} + \dfrac{1-k}{1+k}}{\dfrac{1-k}{1+k}z^{-2} - \dfrac{2\alpha}{1+k}z^{-1} + 1}\right)$ | $\alpha = \dfrac{\cos\left(\dfrac{\omega_2 + \omega_1}{2}\right)T}{\cos\left(\dfrac{\omega_2 - \omega_1}{2}\right)T} = \cos\omega_0 T$  $k = \tan\left(\dfrac{\omega_2 - \omega_1}{2}\right)T \tan\dfrac{\beta T}{2}$ |

## Examples

1. Realize in canonic form the transfer function

$$G(z) = (z^{-1} - \alpha)/(1 - \alpha z^{-1})$$

where $\alpha$ is real and its modulus is less than unity. Find the amplitude and characteristics and comment on their shape.

2. A transfer function has a pair of complex conjugate poles situated at $\sqrt{2}\exp(\pm j\pi/4)$ and a second order zero at $-1$, on the $z^{-1}$-plane. Derive an expression for the transfer function and its amplitude characteristic.

3. A lowpass digital filter is required to have a cutoff frequency of 3 kHz when the sampling frequency is 16 kHz. The passband attenuation is required to be at most 1 dB, whereas the stopband attenuation should be at least 30 dB attained from 4·4 kHz and above. By using appropriate filter tables for continuous filters and employing the bilinear transformation derive an expression for the required digital filter transfer function.

4. In the transfer function derived in Example 3 above, all the delay elements $(z^{-1})$ have been replaced by $(-z^{-1})$. What is the effect of such an operation on the amplitude characteristic?

# Chapter 6

# Filters with Finite Duration Impulse Responses

*G. B. Lockhart*

## 6.1  Introduction

This chapter is the first of several concerned with filters which have finite-duration impulse responses. Such filters include types realized in non-recursive form, frequency sampling filters (throughout Chapter 9) and discrete Fourier transform filters (Chapter 8). The basic design methods to be discussed apply in all cases.

Although provision for more storage and a greater number of arithmetic operations is often necessary in comparison with certain recursive types (Chapter 3), filters with finite-duration impulse responses have a number of advantages. Stability is always ensured and design from an impulse response specification is straightforward. Symmetrical impulse responses and therefore filters with exactly linear phase characteristics are realizable without difficulty. In frequency domain design the discrete Fourier transform relation which exists between the impulse response and sampled values of the frequency response facilitates the application of certain optimization procedures.

## 6.2  Analogue and digital transversal filters

The analogue transversal filter is a well-known device which consists of a tapped delay line, a set of weighting resistors and a summer (Figure 6.1). The output signal is formed by summing the delayed and weighted input signals.

Explicitly, if $x(t)$ is the signal input to an $N$-tap delay line the output signal $y(t)$, is given by,

$$y(t) = \sum_{n=0}^{n+1} h_n x(t - nT) \tag{6.1}$$

where $T$ seconds is the delay between taps and $(h_0, h_1, \ldots, h_{N-1})$ are constant weighting coefficients which determine a particular filtering action. Since delay, multiplication by a constant coefficient and summation are the only operations employed to generate $y(t)$ from $x(t)$ and the filter is linear and therefore, in common with other linear networks, may be characterized by an impulse and frequency response.

A digital realization of the linear transversal filter is feasible if the input signal is first sampled and quantized to a digital form (Figure 6.2). An $N$-stage digital delay line (shift register) is employed which holds successive numbers (usually in a binary format) generated by the analogue to digital converter

76

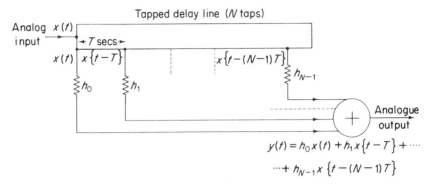

**Figure 6.1** Transversal filter

and shifts the sequence by one stage when a clock pulse is applied every $T$ seconds. The part between X and Y (Figure 6.2) is strictly a digital filter since all operations are performed digitally and both input, $x(kT)$, and output, $y(kT)$, are numerical sequences ($k = \cdots -1, 0, +1 \ldots$). The input sequence may be conceived as a sampled and digitized version of a continuous analogue input although in practice, it need not have been derived in this way.

The value of an output number, $y(kT)$ is related to the input sequence by

$$y(kT) = \sum_{n=0}^{N-1} h_n x[(k-n)T]. \tag{6.2}$$

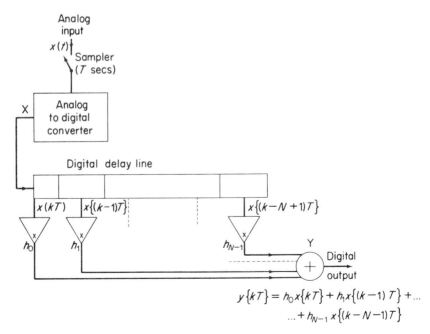

**Figure 6.2** Digital transversal filter

This expression is analogous to the analogue transversal filter input/output relation of equation 6.1. It should be noted, however, that the accuracy to which weighting coefficients can be set in the digital case depends on the number of digits employed and also that multiplication by digital arithmetic will usually involve 'round-off' errors. (Round-off errors can be interpreted as 'noise' and statistical methods employed to analyse their combined effect on the filter performance. Such considerations will be discussed in Chapter 10.)

The digital version of the transversal filter (Figure 6.2) is termed 'non-recursive' since each output number (equation 6.2) depends only on the input sequence. Unlike the recursive digital filter, weighted sums of the output sequence are not involved.

**Example 1.** Consider a non-recursive digital filter which forms output values by taking the difference of the current input value and the previous input delayed by $T$ seconds (Figure 6.3). Since each number in the output sequence is proportional to the rate of change of the input sequence, this filter may be regarded as a crude differentiator.

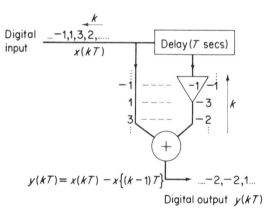

**Figure 6.3**   Simple differentiator

The output is given by,

$$y(kT) = x(kT) - x[(k - 1)T]$$

so that, with reference to equation 6.2

$$h_0 = 1$$
$$h_1 = -1$$
$$h_k = 0, k > 1.$$

The response of the filter to the sequence,

$$\ldots 2, 3, 1, -1, \ldots$$

is illustrated in Figure 6.3.

## 6.3 Impulse response

If the 'impulse' sequence

$$1, 0, 0, 0, \ldots$$

is applied to the input of the general non-recursive digital filter (Figure 6.4) the output sequence will be

$$h_0, h_1, h_2, \ldots, h_{N-1}, 0, \ldots.$$

The impulse response is therefore formed from the set of $N$ weighting constants. Since the number of delay elements must always be finite it follows that the impulse response of the non-recursive filter is always limited in time. This contrasts with the theoretically infinite impulse response of the recursive filter. It also follows that any digital filter which has a finite duration impulse response can be realized in the non-recursive form of Figure 6.4.

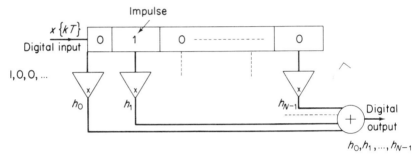

**Figure 6.4**  Response to impulse sequence

**Example 2.** Consider again the crude differentiator of Example 1. If the impulse sequence is applied to the input (Figure 6.5), the output response is,

$$1, -1, 0, 0, \ldots.$$

This sequence is therefore the impulse response of the filter.

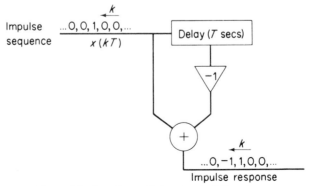

**Figure 6.5**  Impulse applied to simple differentiator

## 6.4 Discrete convolution

Any input sequence to the filter may be regarded as a sum of impulse sequences, each applied at different sampling instances. For example, let the impulse response of a non-recursive filter be:

$$2{\cdot}00,\ 1{\cdot}00,\ 0{\cdot}50,\ 0{\cdot}25,\ 0{\cdot}00,\ 0{\cdot}00,\ \ldots$$

and the input sequence,

$$1{\cdot}00,\ 0{\cdot}50,\ 0{\cdot}50,\ 0{\cdot}00,\ 0{\cdot}00,\ \ldots.$$

The input sequence is represented as three weighted impulse sequences successively applied to the input as illustrated in Figure 6.6. The three corresponding impulse responses are then summed to determine the total response. The output sequence can therefore be regarded as the input sequence 'convolved' with the impulse response.

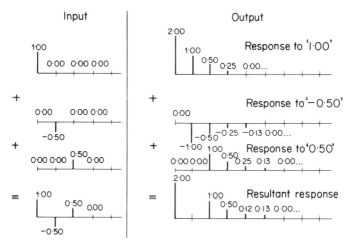

**Figure 6.6**  Illustration of discrete convolution

In general, equation 6.2 is an expression of 'discrete' convolution between the input sequence and impulse response, whereas the corresponding relation for the conventional analogue filter expresses 'continuous' convolution in which the summation of equation 6.2 is replaced by an integral. If transform methods are applied, convolution operations in the time domain may be replaced by multiplication in the frequency domain as the following section will demonstrate for the discrete case.

## 6.5 Frequency response

The output of a non-recursive filter is given by the convolution of the input sequence and the impulse response, i.e. (from equation 6.2),

$$y(kT) = \sum_{n=0}^{N-1} h_n x[(k-n)T]. \tag{6.3}$$

Suppose that the input sequence is a sampled and digitized cosine wave of constant frequency, $\omega$ rad/sec, then,

$$x(kT) = \cos k\omega T, \qquad k = \cdots -1, 0, +1, \ldots.$$

The output sequence, $y_\omega(kT)$, is determined by substituting the above expression for $x(kT)$ in equation 6.3. Thus,

$$y_\omega(kT) = \sum_{n=0}^{N-1} h_n \cos [(k-n)\omega T]$$

$$= \text{Re} \sum_{n=0}^{N-1} h_n \, e^{j(k-n)\omega T}$$

where Re, denotes, 'real part of' and $j = \sqrt{-1}$. Part of the exponential may be taken outside the summation so that

$$y_\omega(kT) = \text{Re} \cdot e^{jk\omega T} \sum_{n=0}^{N-1} h_n \, e^{-jn\omega T}$$

$$= \text{Re} \cdot e^{jk\omega T} H(e^{j\omega T})$$

where

$$H(e^{j\omega T}) = \sum_{n=0}^{N-1} h_n \, \bar{e}^{jn\omega T}. \qquad (6.4)$$

Hence,

$$y_\omega(kT) = \text{Re} \cdot |H(e^{j\omega T})| \, e^{j[k\omega T + \theta(\omega)]}$$

in which,

$H(e^{j\omega T})$ has been expressed as

$H(e^{j\omega T}) = |H(e^{j\omega T})|e^{j\theta(\omega)}.$

Finally, taking the real part,

$$y_\omega(kT) = |H(e^{j\omega T})| \cos [k\omega T + \theta(\omega)].$$

Thus the response of the filter to a cosine wave of frequency $\omega$ is a cosine wave of the same frequency but with values of magnitude and phase determined by $|H(e^{j\omega T})|$ and $\theta(\omega)$ respectively. $H(e^{j\omega T})$ is therefore the characteristic frequency response of the filter. It is apparent from equation 6.4 that $H(z)$ is the $z$-transform of the impulse response $(h_0, h_1, h_2, \ldots, h_{N-1})$. That is,

$$H(z) = \sum_{n=0}^{N-1} h_n z^{-n}. \qquad (6.5)$$

It follows that $H(e^{j\omega T})$ the frequency response of the filter is obtained by taking the $z$-transform of the impulse response, and setting $z = e^{j\omega T}$. Again there is direct analogy with the conventional analogue filter for which the frequency response is obtained by taking the Laplace transform $H(s)$ of the impulse response $h(t)$ and setting $s = j\omega$.

## 6.6  Poles and zeros

In addition to impulse and frequency responses, digital filters may be characterized by the location of the poles and zeros of $H(z)$ the $z$-transform of the impulse response. The right hand side of equation 6.5 is an $N$th degree polynomial in $z^{-1}$ and therefore can be factorized so that,

$$H(z) = h_{N-1}(z^{-1} - \alpha_1)(z^{-1} - \alpha_2)\dots(z^{-1} - \alpha_{N-1})$$

where $(\alpha_1, \alpha_2, \dots, \alpha_{N-1})$ are the zeros of $H(z^{-1})$. They will generally be complex and satisfy the equation,

$$H(z^{-1}) = 0.$$

Because the impulse response $(h_0, h_1, \dots, h_{N-1})$ consists of real numbers zeros are either real or form conjugate pairs. A typical $z^{-1}$-plane zero pattern is illustrated in Figure 6.7. Since the $z$-transform of the non-recursive filter is always a finite degree polynomial (equation 6.5) no poles can appear in any finite part of the $z^{-1}$-plane. It follows that the non-recursive filter is always stable. This of course is consistent with the absence of feedback.

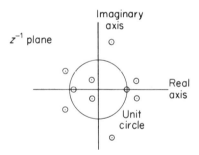

**Figure 6.7**  Characteristic zero pattern of non-recursive filter

**Example 3.** It is known from Example 2 that the impulse response of the simple differentiator (Figure 6.5) is the sequence $1, -1, 0, \dots$. The $z$-transform of the impulse response is therefore given by,

$$H(z) = (1 - z^{-1}).$$

Because $H(1) = 0$, the $z^{-1}$-plane contains a single zero at $z^{-1} = 1$ (Figure 6.8). Setting $z = e^{j\omega T}$ in order to determine the frequency response,

$$H(e^{j\omega T}) = (1 - e^{-j\omega T})$$

82

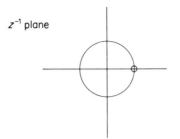

**Figure 6.8** Simple differentiator zero pattern

and

$$|H(e^{j\omega T})| = |1 - e^{j\omega T}|$$

$$= \sqrt{2(1 - \cos \omega T)}$$

$$= 2|\sin (\omega T/2)|.$$

$|H(e^{j\omega T})|$, the magnitude of the frequency response is illustrated in Figure 6.9.

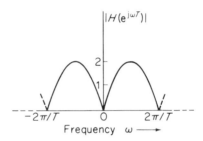

**Figure 6.9** Frequency characteristic of simple differentiator

## 6.7 Analogue and non-recursive digital filters compared

Properties of the non-recursive digital filter which have been discussed are summarized and compared with those of the conventional analogue filter in Table 6.1. An asterisk indicates properties which do not apply in the case of the recursive digital filter.

## 6.8 Coefficient quantization

In the hardware realization of a digital filter the values of weighting coefficients cannot be set on a continuous scale but must be represented by a fixed number of binary digits. Therefore the coefficient values specified by the filter design are usually 'rounded-off' to the nearest value which can be represented. This will result in a filter which will differ to some extent from the original design.

Table 6.1

| Property | Analogue filter | Non-recursive filter |
|---|---|---|
| Type of input and output signal | Continuous | Numerical Sequences |
| Input/output relation. ('$y$' denotes output, '$x$' input, '$h$' impulse response). | $y(t) = \int_0^{\infty} h(T)x(t - T)\,dT$ | $y(kT) = \sum_{n=0}^{N-1} h_n\, x\,[(k - n)T]$ |
| Type of impulse response | Continuous–not truncated | Truncated numerical, sequence |
| Frequency response | $H(j\omega)$, where $H(s)$ is the Laplace transform of $h(t)$ | $H(e^{j\omega T})$, where $H(z)$ is the $z$-transform of $h(nT)$, $n = 0, 1, \ldots, (N - 1)$ |
| Type of transfer function | $H(s)$, rational function in '$s$' | $H(z)$, polynomial in '$z^{-1}$'* |
| Poles and/or zeros | $H(s)$ possesses both in $s$ plane | $H(z^{-1})$ possesses zeros only in finite $z^{-1}$ plane* |
| Stability | No poles in right hand $s$ plane for stability | Always stable* |

Let $h_n$ be the $n$th coefficient of the impulse response of a non-recursive filter. If the quantized coefficients can assume only the values $kE_0$, ($k = 0, 1, 2, \ldots$) where $E_0$ is a constant then the $n$th quantized coefficient can be expressed,

$$(h_n + e_n)$$

where $e_n$ is an error term, satisfying

$$(-E_0/2 < e_n < +E_0/2).$$

Substituting the expression for the quantized coefficient in equation 6.2,

$$y(kT) = \sum_{n=0}^{N-1} (h_n + e_n)x[(k - n)T]$$

$$= \sum_{n=0}^{N-1} h_n x[(k - n)T] + \sum_{n=0}^{N-1} e_n x[(k - n)T]. \qquad (6.6)$$

In the frequency domain,

$$Y(e^{j\omega T}) = X(e^{j\omega T})H(e^{j\omega T}) + X(e^{j\omega T})E(e^{j\omega T}). \qquad (6.7)$$

The error sequence in equation 6.6 can be conceived as the impulse response of an unwanted filter placed in parallel with the ideal. It is apparent from equation 6.7 that $H(e^{j\omega T})$, the ideal frequency response will be distorted by the superposition of $E(e^{j\omega T})$.

The extent to which the ideal response is distorted at any frequency can be accurately obtained in any particular case by evaluating $E(e^{j\omega T})$ from the error sequence in equation 6.6. A useful estimate of the mean-square error $\overline{|E(e^{j\omega T})|^2}$ however, is easily obtained by adopting the statistical approach to quantization[1]. We note that

$$|E(e^{j\omega T})|^2 = E(e^{j\omega T})E^*(e^{j\omega T})$$

where * denotes complex conjugation, and substituting via equation 6.4 for $E(e^{j\omega T})$,

$$|E(e^{j\omega T})|^2 = \sum_{n=0}^{N-1} e_n e^{-nj\omega T} \sum_{n=0}^{N-1} e_n e^{nj\omega T}$$

$$= \sum_{n=0}^{N-1} e_n^2 + \text{cross-terms.}$$

If the usual assumptions[1] are made that successive error samples are uncorrelated then the expected value of the cross-terms will be zero and the mean-square error may be approximated by the expected value,

$$\overline{|E(e^{j\omega T})|^2} = \sum_{n=0}^{N-1} \bar{e}_n^2.$$

If it is also assumed[1] that errors are uniformly distributed in the range $(-E_0/2 < e_n < E_0/2)$ then,

$$\bar{e}_n^2 = E_0^2/12$$

and hence,

$$\overline{|E(e^{-j\omega T})|^2} = NE_0^2/12.$$

The combined frequency response will therefore approximate the ideal with a root-mean-square deviation of,

$$(E_0/2) \cdot \sqrt{N/3}.$$

It must be emphasized that this is a probabilistic measure of deviation which will almost certainly be exceeded in a particular case for some value of $\omega$. Nevertheless, it is a useful guide, is easily calculated and provides an estimate of the maximum attenuation obtainable in stopband regions when the ideal frequency response approaches zero in magnitude.

### 6.9 Design of non-recursive filters by frequency sampling

The frequency sampling method is formulated entirely in the frequency domain and therefore particularly suited to filter design from frequency domain specifications. The frequency response of an $N$-stage non-recursive filter is given by

$$H(e^{j\omega T}) = \sum_{n=0}^{N-1} h_n e^{-jn\omega T}. \tag{6.4}$$

Suppose that this characteristic is sampled at intervals of $1/NT$ Hz along the frequency axis. If the frequency samples are $(H_0, H_1, H_2, \ldots, H_{N-1})$, then

$$H_r = H(e^{j2\pi r/N})$$

$$= \sum_{n=0}^{N-1} h_n e^{-jn2\pi r/N}. \tag{6.8}$$

The above expression (6.8) which relates the impulse response to the set of frequency samples is a discrete Fourier transform. The impulse reponse may be determined from a specification of the real and imaginary parts of the frequency samples (or equivalently magnitude and phase) by application of the inverse discrete Fourier transform,

$$h_n = \frac{1}{N} \sum_{r=0}^{N-1} H_r e^{jr2\pi n/N}.$$

The evaluation of the impulse response if required, may be performed by the use of a fast Fourier transform (F.F.T.) algorithm. (The theory and application of discrete Fourier transforms and the F.F.T. are discussed in detail in Chapter 7.) If there are $N$ real impulse response coefficients to be set as above then considerations of spectral symmetry lead to the conclusion that the complex values of $(N/2)$ points on the frequency response may be independently selected at intervals of $(1/NT)$ Hz up to the 'foldover' frequency of $(1/2T)$ Hz. Such a specification uniquely determines the impulse response of the non-recursive filter; the frequency response is exactly equal to the specification at each of the $N/2$ points.

**Example 4.** A non-recursive filter has 30 taps and a clock rate of 3 kHz. A lowpass filter response is required with cutoff frequency 700 Hz.

Since there are 30 taps, 15 points on the frequency response at intervals of $\frac{1500}{15} = 100$ Hz may be selected up to 100 Hz below the 'foldover' frequency at 1·5 Hz. Since a lowpass response is required with a 700 Hz cutoff, an ideal rectangular characteristic is sampled so that all samples up to and including the 700 Hz sample are set to the same value and the remainder to zero (Figure 6.10). If the inverse D.F.T. of these samples is taken to determine the appropriate values for the 30 weighting coefficients, then the frequency characteristic of the filter will be as illustrated in Figure 6.10. The filter characteristic passes through the 15 specified points and cuts off at 700 Hz required.

Although the frequency characteristic of a fully-specified filter such as above will assume the required values at sampling points, intersample values can deviate unacceptably from the ideal form. For example, the extent of sidelobe ripple in the lowpass characteristic of Figure 6.10 may well be unacceptable. If it is sufficient to specify less than $N$ 'fixed' sample values of the frequency characteristic, the remaining 'free' sample values can be manipulated in a search for a minimum of some filter parameter. These free samples are usually taken in the transition bands and varied to minimize sidelobe ripple in the

86

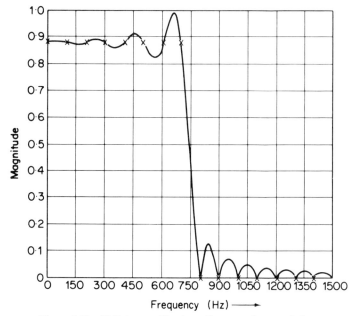

Figure 6.10  15-Point specification of lowpass characteristic

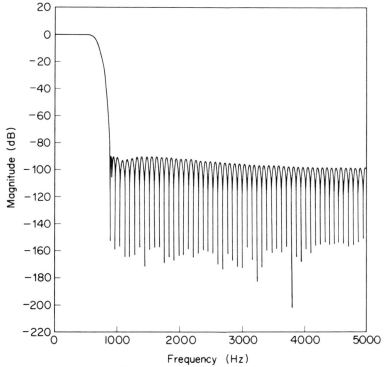

Figure 6.11  Lowpass filter with finite duration impulse response, $N = 128$
(reproduced from Reference 2 by courtesy of IEEE)

stopbands. The allocation of sample points to the transition region naturally increases the transition bandwidth and this is the penalty to be paid for reduced sidelobe ripple. Rabiner *et al.*[2] have utilized an efficient minimization algorithm to determine transition band sample values for a variety of lowpass and bandpass filters, obtaining responses such as those illustrated in Figure 6.11.

### 6.10 Window functions

A common approach to the design of non-recursive filters is based on approximating the infinite duration impulse response of an 'ideal' filter by a finite duration response. Direct truncation of the ideal response is equivalent to multiplication by a rectangular window function and results in the Gibbs phenomenon in the frequency domain with unwanted 'overshoot' at sharp transitions of the ideal response. This effect is also typical of fully-specified filters such as Figure 6.10. Multiplication of an impulse response by a window function corresponds to convolution of the frequency characteristic with the Fourier transform of the window. A careful choice of window function can therefore result in a 'smoothing' of the frequency characteristic with significant reductions in ripple. The properties of a wide variety of window functions have been studied in the literature[2,3,4].

**Example.** Consider the fully-specified filter response of Figure 6.10. The action of a simple window function in reducing sidelobe ripple will be illustrated by multiplying the impulse response by a 'raised cosine' window.

$$[1 + \cos(2\pi t/NT)] = [\tfrac{1}{2} e^{-j2\pi t/NT} + 1 + \tfrac{1}{2} e^{+j2\pi t/NT}].$$

The reference point, $t = 0$, is taken here as the midpoint of the finite duration impulse response so that a symmetrical window is imposed. In the frequency domain the result is the convolution of the response, $H(e^{j\omega t})$ (Figure 6.10) with the Fourier transform of the window which in this case is simply a three-component line spectrum with weights $(\tfrac{1}{2}, 1, \tfrac{1}{2})$ at frequencies

$$\left(-\frac{1}{NT}, 0, +\frac{1}{NT}\right)$$

respectively. The derived response $|H_s(e^{j\omega T})|$ is illustrated in Figure 6.12 and is given by

$$H_s(e^{j\omega T}) = \tfrac{1}{2} H[e^{j(\omega - 2\pi/NT)T}] + H(e^{j\omega T}) + \tfrac{1}{2} H[e^{j(\omega + 2\pi/NT)T}].$$

It is apparent that the original response of Figure 6.10 has been smoothed in such a manner that sidelobe ripple is considerably reduced although the transition bandwidth has been doubled.

### 6.11 Realization

Techniques for the realization of filters with finite duration impulse responses fall into three main categories:

88

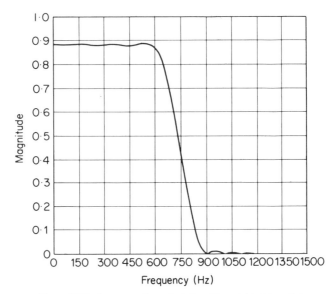

**Figure 6.12**  Lowpass characteristic with weighted impulse
response

(1) *Direct convolution.* The impulse response is determined and the filter realized by the direct computation of equation 6.2.

(2) *Fast convolution.* Convolution is performed by taking the inverse Fourier transform of the product of the discrete Fourier transform of the impulse response and the discrete Fourier transform of a segment of input data. A fast Fourier transform algorithm is employed for efficient computation of the discrete transforms (Chapter 7).

(3) *Frequency sampling.* Frequency samples serve as weights for the outputs of a set of parallel digital resonators (Chapter 8).

In conclusion it is emphasized that the basic characteristics and design problems for all digital filters with finite duration impulse responses are the same regardless of the particular means chosen for realization.

### References

1. Bennet, W. R., 'Spectra of quantised signals', *Bell System Technical Journal*, **27**, 446–472 (1948).
2. Rabiner, L. R., Gold, B., and McGonegal, C. A., 'An approach to the approximation problem for nonrecursive digital filters', *IEEE Trans. on Audio and Electroacoustics*, **AU-18** No. 2, 83–106 (1970).
3. Kaiser, J. F., 'Digital filters,' in *System Analysis by Digital Computers* (Ed. F. F. Kuo, and J. F. Kaiser), Wiley, New York, 1966, Chap. 7.
4. Helms, H. D., 'Nonrecursive digital filters: design methods for achieving specifications on frequency response', *IEEE Trans. on Audio and Electroacoustics*, **AU-16**, 336–342 (1968).

**Examples**

1. The output of a non-recursive filter is formed by taking the average of the current input value and the previous input value. If the clock period is $100\,\mu s$ determine:
   (a) An expression relating output to input.
   (b) Impulse response.
   (c) Frequency response–sketch magnitude and phase functions and compare with the simple differentiator discussed in the chapter.
   (d) The location of zeros in the $z^1$ plane–sketch.
2. If the input to the filter of question 1 is,
   (a) the sequence $\ldots, 1, 1, 1, \ldots$
   (b) a sampled, digitized sinusoid of frequency $5\,kHz$ and peak to peak value 2,
   determine the output sequence directly from the frequency characteristic obtained for question 1(c). Verify by calculating a number of output values by application of the input/output relation [question 1(a)].

# Chapter 7

# Fourier Transform Methods

*R. Coates*

## 7.1 The Discrete Fourier Transform

The Discrete Fourier Transform specifies the line spectrum of a sampled periodic time function. The Inverse Discrete Fourier Transform permits reconstruction of the time function from its spectrum. The two transforms are normally referred to by the abbreviations DFT and IDFT, respectively.

The DFT analyses periodic functions, and its derivation may proceed from a statement of the Fourier Series identities[1]. Let $x_0(t)$ be a continuous periodic function of period $P$ and frequency $f_0 = 1/P$ such that:

$$x_0(t) = x_0(t + mP); \qquad m \text{ integer.}$$

We may represent $x_0(t)$ as a Fourier Series:

$$x_0(t) = \sum_{n=-\infty}^{+\infty} X_0(n) . \exp(2\pi j n f_0 t); \qquad 0 < t < P \tag{7.1}$$

where the Fourier Coefficients $X_0(n)$ are given by:

$$X_0(n) = \frac{1}{P} \int_0^P x_0(t) . \exp(-2\pi j n f_0 t) . dt. \tag{7.2}$$

Normally $x_0(t)$ would be a real function and $X_0(n)$ complex, although neither need be so restricted. Since we have represented $x_0$ as a function of time, the $X_0(n)$ may be interpreted as the complex spectrum of $x_0(t)$. From the real and imaginary parts of $X_0(n)$ may be deduced the amplitude and phase of the components forming $x_0(t)$.

We now examine the effect of sampling on the periodic function $x_0(t)$. In order that the function may be sampled unambiguously it is necessary that its spectrum contains no components beyond some frequency $f_1$. That is:

$$X_0(n) = 0; \qquad |n| > n_1$$

where $n_1$ is the integer value of $n$ defining the frequency $f_1$:

$$n_1 f_0 = f_1.$$

Figure (7.1) illustrates such a spectrum and the waveform from which it is derived.

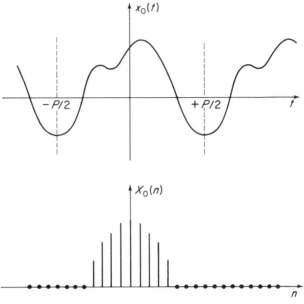

**Figure 7.1**   The periodic baseband function $x_0(t)$ and its spectrum $X_0(n)$

If we apply the Sampling Theorem, which has been discussed in Chapter 2, then the sampling interval, $T$, will be given as:

$$T = \frac{1}{2f_1} = \frac{1}{2n_1 f_0} = \frac{P}{2n_1}$$

and, allowing $N$ samples within the period:

$$2n_1 = N.$$

The sampling process establishes a periodic waveform, normalized with respect to $T$ such that:

$$x(t/T) = \sum_{k=0}^{N-1} x_0(t/T) \cdot \delta(t/T - k) \tag{7.3}$$

and this waveform is defined over its period, so that:

$$0 \leqslant t < P$$

or

$$0 \leqslant t/T < N.$$

Since $x(t/T)$ is periodic, its Fourier Coefficients are evaluated by applying equation (7.2) so that:

$$X(n) = \frac{1}{N} \int_0^{N-1} x(t/T) \cdot \exp\left(\frac{-2\pi jn}{N} \cdot \frac{t}{T}\right) \cdot \mathrm{d}(t/T).$$

(The limits and divisor are changed from $P$ to $N$ to accommodate the normalization of the independent variable.)

Inserting equation (7.3), we obtain:

$$X(n) = \frac{1}{N} \int_0^{N-1} \sum_{k=0}^{N-1} x_0(t/T) . \delta(t/T - k) . \exp\left(\frac{-2\pi jn}{N} . \frac{t}{T}\right) . d(t/T).$$

Now note that the sampling $\delta$-function is so defined that:

$$\int_{-\infty}^{+\infty} f(t/T) . \delta(t/T - k) . d(t/T) = f(k)$$

so that:

$$X(n) = \frac{1}{N} \sum_{k=0}^{N-1} x_0(k) . \exp\left(\frac{-2\pi jkn}{N}\right).$$

Finally, note that, by definition:

$$x(k) = x_0(k)$$

and hence:

$$X(n) = \frac{1}{N} \sum_{k=0}^{N-1} x(k) . \exp\left(\frac{-2\pi jkn}{N}\right). \tag{7.4}$$

The relationship establishing $x(k)$ from $X(n)$ may be immediately written down from equation (7.1) by letting $t = kT$ and noting that the summation conditions are, as a consequence of the bandlimiting of $x_0(t)$, finite. Then:

$$x(k) = \sum_{k=0}^{N-1} X(n) . \exp\left(\frac{2\pi jkn}{N}\right). \tag{7.5}$$

It should be noted that $x(k)$ is really a periodic function, with:

$$x(k) = x(k + mN); \qquad m \text{ integer}$$

and, likewise:

$$X(n) = X(n + mN); \qquad m \text{ integer}$$

That the spectrum should be periodic follows from the periodic nature of the spectrum of any sampled function. Its discrete nature follows from the initial postulate that the function being sampled is, also, periodic.

We conclude that, given a periodic function $x_0$, it may be sampled and the relationship (7.4) applied to the sample values to derive values of a spectrum $X(n)$ which is, within the period $0 \leqslant n \leqslant N - 1$, identical to the spectrum $X_0(n)$ of the original periodic function. Typical waveforms are illustrated in Figure 7.2. Since equation (7.4) was obtained by means of the Sampling Theorem, it provides an exact and economical alternative to the initial integral relation and would therefore be suitable for the machine evaluation of Fourier Coefficients. Equations (7.4) and (7.5), we shall refer to as the Discrete Fourier

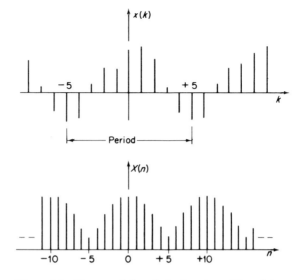

**Figure 7.2** The sampled periodic function $x(k)$ and its periodic spectrum $X(n)$

Transform (DFT) and Inverse Discrete Fourier Transform (IDFT), respectively. Note that now the variable $n$ runs from zero to $N - 1$. We interpret the spectrum in the following way. The first $(N/2 - 1)$ points of $X(n)$ correspond to the $(N/2 - 1)$ positive frequency spectral lines of $X_0(n)$, as shown in Figure 7.3. The last $(N/2 - 1)$ points of $X(n)$ correspond to the $(N/2 - 1)$ negative spectral lines of $X_0(n)$.

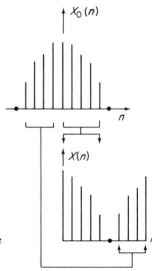

**Figure 7.3** Relation between Fourier series coefficients and the DFT

The transform pair given by equations (7.4) and (7.5) may appear in several forms. For example, the $1/N$ multiplier and the negative exponent may be switched between the pairs without altering their validity as a transform pair. The spectrum may not then be directly identified with that given by equation (7.2), of course. Sometimes both equations are encountered multiplied by $(1/N)^{\frac{1}{2}}$.

## 7.2 Theorems and properties of the discrete Fourier transform

### 7.2.1 DFT in one dimension

It is customary to abbreviate the exponent in the DFT and IDFT equations by writing* :

$$\exp(2\pi j/N) = W.$$

Then:

$$x(k) = \sum_{n=0}^{N-1} X(n)W^{nk}; \qquad k = 0, 1, \ldots, N-1$$

and:

$$X(n) = \frac{1}{N} \sum_{k=0}^{N-1} x(k) . W^{-nk}; \qquad n = 0, 1, \ldots, N-1.$$

### 7.2.2 DFT in r dimensions

The DFT, like the Fourier Transform, may be generalized to $r$ dimensions:

$$x(k_1, k_2, \ldots, k_r) = \sum_{n_1=0}^{N_1-1} \sum_{n_2=0}^{N_2-1} \cdots \sum_{n_r=0}^{N_r-1} W_{N_1}^{n_1 k_1} . W_{N_2}^{n_2 k_2} \ldots W_{N_r}^{n_r k_r} . X(n_1, n_2, \ldots, n_r)$$

$$X(n_1, n_2, \ldots, n_r) = \frac{1}{\prod_{m=1}^{r} N_m} . \sum_{k_1=0}^{N_1-1} \sum_{k_2=0}^{N_2-1} \vdots \sum_{k_r=0}^{N_r-1} W_{N_1}^{-n_1 k_1} . W_{N_2}^{-n_2 k_2} \ldots W_{N_r}^{-n_r k_r}$$

$$\times x(k_1, k_2, \ldots, k_r)$$

with $k_s, n_s = 0, 1, \ldots N_s - 1$ for $s = 1, 2, \ldots, r$.

### 7.2.3 Use of the DFT in forming the IDFT

We may, by means of a simple data rearrangement and scaling, use the DFT equation to form the IDFT. This is of importance in that it allows one algorithm to be used for performing both transforms.

We expand the DFT to give:

$$X(n) = [x(0) . W^0 + x(1) . W^{-1.n} + \cdots + x(N-1) . W^{-(N-1).n}]/N$$

* Note, however that some authors use the negative exponent!

and reverse the order:

$$X(n) = [x(N-1) \cdot W^{-(N-1)n} + x(N-2) \cdot W^{-(N-2)n} + \cdots + x(0) \cdot W^0]/N$$

$$= [x(N-1) \cdot W^{-Nn} \cdot W^{+n} + x(N-2) \cdot W^{-Nn} \cdot W^{+2n} + \cdots$$

$$+ x(0) \cdot W^{-Nn} \cdot W^{+Nn}]/N$$

but note that:

$$W^{-Nn} = \exp(-2\pi jn)$$

$$= 1, n \text{ integer.}$$

Then:

$$X(n) = [x(0) \cdot W^0 + x(N-1) \cdot W^{+n} + \cdots + x(1) \cdot W^{+(N-1)n}]/N$$

which has the IDFT form.

We deduce that, to perform an IDFT on an array $X(n)$ we leave $X(0)$ alone and reverse the remaining sequence of numbers, replacing $X(1)$ with $X(N-1)$, $X(2)$ with $X(N-2), \ldots$ and vice versa. Finally all the rearranged numbers are multiplied by $N$.

### 7.2.4 Parseval's Theorem

Given two time functions $x(k)$ and $y(k)$ with DFT's $X(n)$ and $Y(n)$, and noting that:

$$y(k) = \sum_{n=0}^{N-1} Y(n) \cdot W^{nk} \qquad k = 0, 1, \ldots, N-1$$

then:

$$y^*(k) = \sum_{n=0}^{N-1} Y^*(n) \cdot W^{-nk}.$$

It follows that the mean value of the product of the two sequences $x(k)$ and $y^*(k)$ is:

$$\frac{1}{N} \sum_{k=0}^{N-1} x(k) \cdot y^*(k) = \frac{1}{N} \sum_{k=0}^{N-1} x(k) \sum_{n=0}^{N-1} Y^*(n) \cdot W^{-nk}.$$

Reversing the order of summation, we obtain:

$$\frac{1}{N} \sum_{n=0}^{N-1} Y^*(n) \sum_{k=0}^{N-1} x(k) \cdot W^{-nk} = \sum_{n=0}^{N-1} Y^*(n) \cdot X(n).$$

Thus we obtain the general result:

$$\frac{1}{N} \sum_{k=0}^{N-1} x(k) \cdot y^*(k) = \sum_{n=0}^{N-1} X(n) \cdot Y^*(n). \tag{7.6}$$

In particular, if we set $y(k) = x(k)$ then, since $x(k) \cdot x^*(k) = |x(k)|^2$ it follows that:

$$\frac{1}{N} \sum_{k=0}^{N-1} |x(k)|^2 = \sum_{n=0}^{N-1} |X(n)|^2. \tag{7.7}$$

This result is known as Parseval's Theorem. It shows that the average 'power' in the sampled time function is equal to the sum of the powers associated with the individual Fourier components of the function, and is unaffected by their phase relationship.

### 7.2.5 Orthogonality

The orthogonality property of discrete cisoid series may be stated thus:

$$\sum_{m=0}^{N-1} W^{mn} W^{mp} = N \quad \text{if } n(\text{mod } N) = -p \tag{7.8}$$

$$= 0 \quad \text{if } n(\text{mod } N) \neq -p.$$

A statement of the form:

$$n(\text{mod } N) = a$$

simply means that $n$ is such that, when it is repeatedly divided by $N$ it leaves a remainder equal to $a$. For example, $12(\text{mod } 5) = 2$. This is equivalent to demanding that:

$$n = \alpha N + a; \quad \alpha \text{ integer.}$$

The first condition is easily proved, since if $n(\text{mod } N) = -p$, then $n = \alpha N - p$ and

$$W^{mn} W^{mp} = W^{m\alpha N}.$$

Now, each term of the summation is of the form:

$$W^{m\alpha N} = 1$$

so that the summation over $N$ terms is seen to equal $N$.

To demonstrate the second condition, it is only necessary to note that

$$W^{mn} W^{mp} = W^{m\beta}$$

where $\beta$ is integer, but non-zero and not equal to an integer multiple of $N$. Then, firstly:

$$W^{m\beta} \neq 1.$$

Secondly, $W^m$ is the $m$th of $N$ roots of unity. Vectorially the $N$ roots of unity are illustrated in Figure 7.4, for $N = 8$. Multiplication of the index $m$ by $\beta$ still results in the same vector diagram. The polygon which is constructed by a graphical execution of the summation is $N$-sided and regular provided $\beta$ is not an integer multiple of $N$. The tail of the first vector is thus joined by the head of the last, so that the vector sum is zero.

98

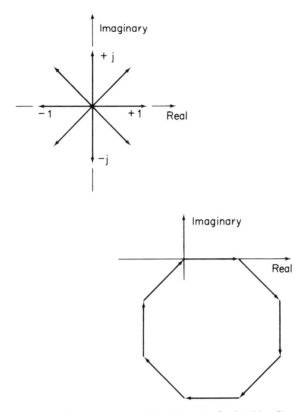

**Figure 7.4** Vector sum of the $N$th roots of unity $(N = 8)$

### 7.2.6 Power spectrum

The DFT of a periodic time-series $x(k)$ of period $N$ is its complex spectrum $X(n)$. It is often very helpful to examine not $X(n)$ but rather the 'power spectrum' of $x(k)$, which corresponds to the 'power' associated with each component of the complex line spectrum. The 'power' contained in the $n$th component of the spectrum is given in the normal way by the product

$$P_{xx}(n) = X(n) \cdot X^*(n)$$
$$= |X(n)|^2; \qquad n = 0, 1, \ldots, N - 1.$$

(7.9)

Strictly speaking, this quantity may be dimensionally only proportional to a dissipated power. For example, if $x(k)$ were measured in volts, $X(n)$ would also be measured in volts, and $P_{xx}(n)$ in (volt)$^2$ rather than watts. It is common convention, nontheless, to refer to this product as a power.

If $x(k)$ had been derived from a continuous periodic time function $x(t)$, then provided $x(t)$ were suitably bandlimited so that $X(n)$ did not exhibit the effects of aliasing, $P_{xx}(n)$ would be an exact periodic repetition of the power spectrum of $x(t)$.

The most important application of power spectral analysis is in the investigation of continuous aperiodic waveforms. In this case, a complex continuous frequency spectrum may not have much meaning in an interpretive sense. Unfortunately, merely segmenting out and sampling a section of such a waveform may well be quite inadequate in approximating the spectrum. The principal reason for this is that the segment may not be truly representative of the whole. Such a procedure leads to a 'raw' or 'unsmoothed' power spectrum. Smoothing can be obtained by an interpolation performed on the $P_{xx}(n)$ and this topic is discussed at length in Section 7.11.

### 7.2.7 Autocorrelation function

The autocorrelation function (ACF) $R_{xx}(k)$, is given by the relation

$$R_{xx}(k) = \frac{1}{N} \cdot \sum_{m=0}^{N-1} x(m)x(k+m); \qquad k = 0, 1, \ldots, N-1. \tag{7.10}$$

Recall that the autocorrelation function of a continuous variable and its power spectrum form a Fourier Transform pair. The analogous relationship for the discrete variable is:

$$R_{xx}(k) = \sum_{n=0}^{N-1} P_{xx}(n)W^{kn}.$$

This we may prove by inserting transform identities into the defining equation for $R_{xx}(k)$ thus:

$$R_{xx}(k) = \frac{1}{N} \sum_{m=0}^{N-1} \left[ \sum_{p=0}^{N-1} X(p)W^{pm} \right]\left[ \sum_{n=0}^{N-1} X(n)W^{kn}W^{nm} \right].$$

Exchanging summations, we obtain:

$$R_{xx}(k) = \frac{1}{N} \sum_{n=0}^{N-1}\sum_{p=0}^{N-1} X(n)X(p)W^{kn}\left[ \sum_{m=0}^{N-1} W^{m(n+p)} \right].$$

The orthogonality property of cisoid summations (equation 7.8), shows that the summation over $m$ is, within the range of interest, of value:

$$N \text{ if } p = -n$$

$$0 \text{ otherwise.}$$

Thus

$$R_{xx}(k) = \sum_{n=0}^{N-1} X(n)X(-n)W^{kn}.$$

However, the DFT has the complex conjugate symmetry property

$$X(-n) = X^*(n)$$

so that:

$$R_{xx}(k) = \sum_{n=0}^{N-1} X(n)X^*(n)W^{kn}$$

$$= \sum_{n=0}^{N-1} P_{xx}(n)W^{kn}. \qquad (7.11)$$

### 7.2.8 Cross-power spectrum and crosscorrelation function

The results of the previous two sections may be generalized. Given two functions $x(k)$ and $y(k)$, the cross-power spectrum is given by

$$P_{xy}(n) = X(n) \cdot Y^*(n)$$

$$= X^*(n) \cdot Y(n); \qquad n = 0, 1, \ldots, N - 1$$

and the crosscorrelation function by:

$$R_{xy}(k) = \frac{1}{N} \sum_{m=0}^{N-1} x(m) \cdot y(k + m)$$

$$= \frac{1}{N} \sum_{m=0}^{N-1} x(k + m) \cdot y(m)$$

$$= \sum_{n=0}^{N-1} P_{xy}(n) \cdot W^{nk} \qquad (7.12)$$

all for $k = 0, 1, \ldots, N - 1$. Again, if $x$ and $y$ are not periodic, smoothing may be necessary.

### 7.2.9 Convolution

The operation of convolution of two periodic functions $x(k)$ and $h(k)$ is defined by the equation (Chapter 2):

$$y(m) = \sum_{i=0}^{N-1} x(i)h(m - i); \qquad m = 0, 1, \ldots, N - 1. \qquad (7.13)$$

There is a considerable similarity between this expression and that used to obtain the correlation function, Section 7.2.7. The only important difference is the 'reversal' of the sequence $h$ prior to its term-by-term multiplication with $x$. Noting this reversal (the negation of the variable of summation, $i$) and following through exactly the same analysis as was used to relate the ACF and the power spectrum, we arrive at the expression:

$$y(m) = \sum_{n=0}^{N-1} [X(n)H(n)]W^{nm} \qquad m = 0, 1, \ldots, N - 1. \qquad (7.14)$$

We see, then, that convolution, a 'time domain' operation, is directly equivalent to multiplication in the 'frequency domain'.

Both these equations are of great importance in that they allow us to perform linear processing of signals and to simulate linear systems. For these classes of application we regard $x(k)$ as the system input, $y(m)$ as the system output and $h(k)$ as the system impulse response.

We may, then, either apply the convolution directly, implementing the first equation, or indirectly by calling upon the DFT to allow us to transform our periodic time functions into the frequency domain. In the latter case we should specify $h(k)$ and $x(k)$ and calculate transforms $H(n)$ and $X(n)$, respectively. These would then be multiplied to yield $Y(n)$. Finally, $Y(n)$ would be retransformed by means of the IDFT to give $y(m)$, the system output.

At first sight, performing the convolution operation in the frequency domain would seem to be a very long-winded approach to a basically straightforward computing task. In fact, this indirect method can, under some circumstances, save a considerable amount of computing time. We shall see why this is so in Section 7.10 after we have discussed a particularly efficient method of computing the DFT and IDFT, known as the Fast Fourier Transform.

Another objection to the method outlined above is that it bears only a superficial resemblance to the non-recursive digital filter equation:

$$y(m) = \sum_{i=0}^{L-1} x(i)h(m-i); \qquad \text{all } m.$$

Note that the impulse response is only $L$-points long. Note also that the lengths of $x(k)$ and $y(k)$ are unrestricted. We are not processing a periodic function with a periodic impulse response in this case. The problem may be resolved, however, by resorting to some rather ingenious data manipulations. These we shall consider at length in Section 7.10.

## 7.3 The analysis of continuous systems

We have seen that the DFT is specifically concerned with the analysis and processing of discrete, periodic signals. It is tempting to apply the DFT directly to provide a numerical analysis of sampled versions of continuous signals*. Thus it would at first sight seem to be possible to establish, by applying either the DFT or the IDFT, discrete values of the Fourier Transform:

$$G_0(j\omega) = \int_{-\infty}^{+\infty} g_0(t) \exp(-j\omega t) \, . \, dt$$

or its inversion integral:

$$g_0(t) = \frac{1}{2\pi} \int_{-\infty}^{+\infty} G_0(j\omega) \exp(j\omega t) \, . \, d\omega.$$

Unfortunately, for reasons which we shall discuss shortly, many signals do not permit an exact correspondence in this way between the Fourier Transform and the DFT. Consequently, it is important that we develop a rationale for handling continuous signals in general.

---

* For one very restricted set of signals this would be a perfectly valid application. The set in question consists of periodic, band-limited waveforms sampled in accordance with the sampling theorem.

Assuming that a continuous signal is not periodic and bandlimited, it may be categorized by the following set of tests. It may be: (a) zero-valued in the time domain beyond certain limits, or (b) zero-valued in the frequency domain beyond certain limits. Note that these properties are mutually exclusive. Neither or either may occur, but not both.

If the signal $g_0(t)$ is zero-valued in the time domain beyond certain limits, then its spectrum $G_0(j\omega)$ cannot be bandlimited. If $G_0(j\omega)$ is of 'lowpass' form then we should expect that $g_0(t)$ could be sampled at a frequency $f_s$ which was sufficiently high to produce a wave $g(t)$ with a spectrum $G(j\omega)$ which would have a reasonable correspondence with $G_0(j\omega)$ over the frequency range

$$-f_s/2 \leqslant f \leqslant + f_s/2.$$

This situation is illustrated in Figure 7.5. Clearly, it is necessary to select the sampling frequency $f_s$ so that the ratio:

$$\frac{|G(2\pi j(f_s/2))|}{|G_0(0)|} \tag{7.15}$$

is 'small enough' to cause no trouble in interpreting the spectrum $G(j\omega)$.

Although it is not possible to lay down a general rule whereby the sampling frequency may be specified for a bandlimited function, some guidelines may be provided. Suppose that the waveform has a lowpass spectrum falling off

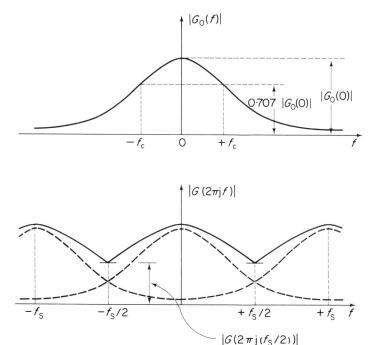

**Figure 7.5** Aliasing of spectrum caused by the sampling process; non-bandlimited waveform

asymptotically from its nominal cutoff frequency (which for convenience we shall regard as the point of intersection of the asymptote and the frequency axis) at $20n$ dB decade$^{-1}$. Such a spectrum, which we shall refer to as of '$n$th order', is illustrated in Figure 7.6. Then the curves given in Figure 7.7 specify the ratio given above, equation (7.14), as the ratio between sampling frequency and nominal cutoff frequency varies.

The second category of signals implies that $g_0(t)$ is bandlimited:

$$G_0(j\omega) = 0; \qquad f > f_1.$$

Since this is the case, the waveform must exist for all time. We suppose also that $g_0(t)$ is aperiodic. It might, for example, be a noise wave subject to ideal rectangular lowpass filtering. The extraction of a segment of duration $T$ containing

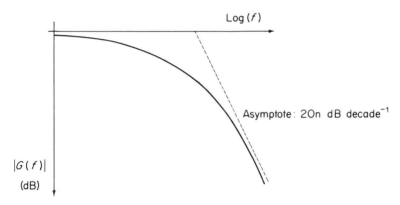

**Figure 7.6**  The $n$th order baseband signal spectrum

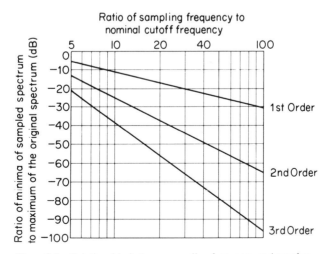

**Figure 7.7**  Relationship between sampling frequency and overlap of aliased spectra for non-bandlimited lowpass signals

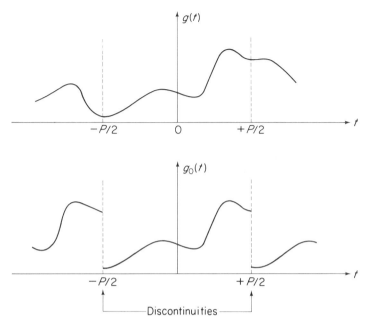

**Figure 7.8** Discontinuities caused by waveform segmentation

$N$ samples is illustrated in Figure 7.8. The DFT process effectively establishes a periodic repetition of the segment extracted from $g(t)$. If $N$ is not large, then the possible discontinuities at the ends of the segment may cause gross errors when the DFT of the derived sequence $x(k)$ is compared with the Fourier Transform of $g(t)$. This leads us to consider two aspects of the DFT process. Firstly, if filtering is to be carried out (that is, evaluating the $X(n)$, weighting them according to some particular law, and returning via the IDFT to the time domain) then care will have to be exercised in choosing $N$ so that a meaningful weighting and an accurate return can be made. Secondly, if a smoothed power spectrum is required, again $N$ must be large, and it may be necessary to preweight the sampled data to remove discontinuities at the ends of the data segment.

### 7.4 Computing the DFT

In Section 7.2.3 it was shown that the DFT and IDFT equations may both be performed by means of a single algorithm. That is, given a routine by means of which we can compute either transform, a simple data rearrangement is all that is needed to allow us to compute the other. Consequently, we need only consider the development of a basic routine for performing a calculation:

$$A(n) \equiv \sum_{k=0}^{N-1} a(k)W^{-nk}; \qquad k = 0, 1, \ldots, N - 1. \tag{7.16}$$

From such an algorithm, we should evaluate the DFT by equating variables in the following way. On input, we should let

$$a(k) \equiv x(k); \qquad k = 0, 1, \ldots, N - 1 \qquad (7.17)$$

so that on output:

$$\frac{A(n)}{N} \equiv X(n); \qquad n = 0, 1, \ldots, N - 1. \qquad (7.18)$$

To use the same algorithm to perform the IDFT, the data to be retransformed would be applied in the following manner:

$$a(k) \equiv X(N - k); \qquad n = 1, 2, \ldots, N - 1$$
$$a(0) \equiv X(0) \qquad (7.19)$$

and would yield a result equivalent to the required time domain sequence:

$$A(n) \equiv x(n); \qquad n = 0, 1, \ldots, N - 1. \qquad (7.20)$$

It is, of course, very easy to derive an algorithm which would allow us to compute the $N$ values of $A(n)$ from the $N$ values of $a(k)$, as dictated by equation (7.15). An example is illustrated in the flow diagram shown in Figure 7.9. Indeed, writing a suitable program from this flow diagram is so simple that it disguises the fact that there are many redundant operations in such a computation. These redundant operations arise in the following way. The general product in equation (7.15) is:

$$a(k)W^{-nk}.$$

The weighting function, $W$, is periodic, of period $N$. For a given value of $k$, as $n$ increases from zero to $N - 1$, the product will also exhibit periodicity and will cycle over to $k$ periods. Furthermore, even over one period, the product may exhibit complex conjugate symmetry. Consequently, with careful tabulation of intermediate results, it is possible to substantially reduce the number of multiplications required to calculate the $N$ values of the sequence $A(n)$.

Algorithms which achieve this reduction in computational work-load are collectively known as 'Fast Fourier Transforms', (FFT). It should be noted that they perform the DFT and are not a 'new' transform.

Another, more modest improvement in efficiency can be obtained when the data to be transformed are real. This is because the DFT of real data also exhibits complex conjugate symmetry. Hence we need only compute one half of the spectrum. The details of this process are presented in Section 7.9.

The DFT is now finding application in so many diverse fields[2] that, in addition to its software implementation, special purpose hardware has been developed. This generally falls into one of two categories: the FFT is performed either by a peripheral unit attached to a large computer, or by a special purpose machine intended only for frequency domain analysis and processing. The art is quite well developed; Reference 3 lists some 200 machines capable of performing the FFT.

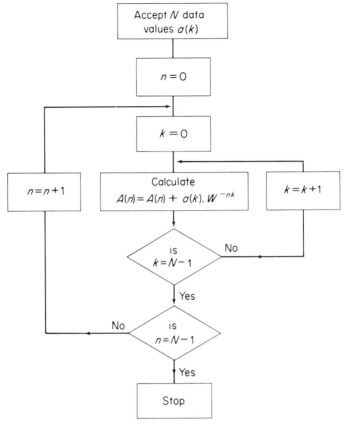

**Figure 7.9** A straightforward implementation of the DFT

'Microprogramming' offers the possibility of obtaining the advantages of a hard-wired FFT with relatively little development labour. Microprogramming is a technique in which a software routine is embedded in a microcircuit 'programmable read-only memory', or PROM. The usual course of implementation is to obtain an algorithm and from this, a software programme which is converted into machine instructions in binary form. These instructions are read onto the PROM, turning it into a permanently programmed replacement for that small section of the computer memory which would normally contain, after compilation, the binary version of the software routine.

## 7.5 Fast Fourier transform

A number of different explanations[4,5,6,7] of the development of the FFT from the DFT have been presented by students of the subject. For example, the DFT may be regarded as a matrix manipulation and the FFT as the result of a matrix factorization process. Concise recursive FFT equations have also been established. These are of great value in that they mathematically formalize the Fast Transform operation. They are also most useful if computer programs are to be written.

For our purposes, a less rigorous approach is sufficient to explain the derivation, performance and programming of the FFT. We take as our starting point equation (7.15), which may be written as:

$$A(n) = \sum_{k=0}^{N-1} a(k)\exp(-2\pi jkn); \qquad n = 0, 1, \ldots, N - 1.$$

Consider the sequence $a(k)$, illustrated in Figure 7.10: split this into two subsequences $y(k)$ and $z(k)$, so that

$$\left.\begin{array}{l} y(k) = a(2k) \\ z(k) = a(2k + 1) \end{array}\right\} k = 0, 1, \ldots, \left(\frac{N}{2} - 1\right).$$

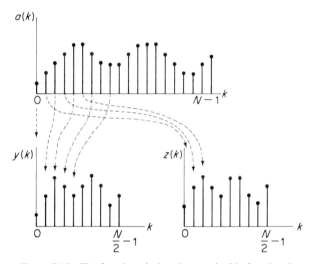

**Figure 7.10**   The first data decimation required in forming the FFT

Let the DFT of the $y(k)$ and $z(k)$ be $Y(n)$ and $Z(n)$, respectively. Then we may write $A(n)$ as:

$$A(n) = \frac{1}{N} \sum_{k=0}^{N/2-1} \left\{ a(2k)\exp\left(\frac{-2\pi j . 2k . n}{N}\right) \right.$$

$$\left. + a(2k + 1)\exp\left[\frac{-2\pi j . (2k + 1) . n}{N}\right]\right\}$$

$$= \frac{1}{2}\left[Y(n) + \exp\left(\frac{-2\pi jn}{N}\right)Z(n)\right].$$

Or, using our more compact notation for the exponential function:

$$A(n) = \tfrac{1}{2}[Y(n) + W^n Z(n)]. \tag{7.21}$$

From a computational standpoint, we should think of $A(n)$ as a complex array of dimension $N$. Likewise, $Y(n)$ and $Z(n)$ would also be complex arrays. Since both of these functions are periodic, of period $N/2$, the arrays containing them need only be of dimension $N/2$. Indeed we may very conveniently employ the same storage locations to hold both $A(n)$ and the two smaller arrays $Y(n)$ and $Z(n)$, thus:

$$\left. \begin{aligned} A(n) &\equiv Y(n) \\ A(n + N/2) &\equiv Z(n) \end{aligned} \right\} n = 0, 1, \ldots, \left(\frac{N}{2} - 1\right).$$

Then the computation to obtain the final transform values would be defined as:

$$A(n) \leftarrow \tfrac{1}{2}\{A(n) + A(n + N/2)W^n\}.$$

Of course, if we choose to specify the computation in this way, we find that we cannot calculate values of $A(n)$ for $n$ greater than $(N/2 - 1)$. However, $W^n$ is a periodic function, of period $N$, with the property that:

$$W^{n+N/2} = -W^n.$$

Hence two computational operations will certainly permit the evaluation of all the $A(n)$:

$$A(n) \leftarrow \tfrac{1}{2}\{A(n) + A(n + N/2)W^n\}$$

$$A(n + N/2) \leftarrow \tfrac{1}{2}\{A(n) - A(n + N/2)W^n\}$$

both for $n = 0, 1, \ldots, (N/2 - 1)$. Consequently, we may compute all other values of $A(n)$ by performing two $N/2$-point transforms, followed by a weighting operation. This process is illustrated in Figure 7.11.

We may evaluate the two $N/2$-point transforms by using exactly the same trick, providing that we again decimate each input sequence into two new subsequences of length $N/4$. We must also suitably adjust the weighting function in the two operation statements. That is, the initial pair was devised to correspond to an $N$-point transform for which:

$$W_N = \exp(-2\pi j/N)$$

For the $N/2$-point transforms:

$$W_{N/2} = \exp[-2\pi j/(N/2)]$$

$$= W_N^2.$$

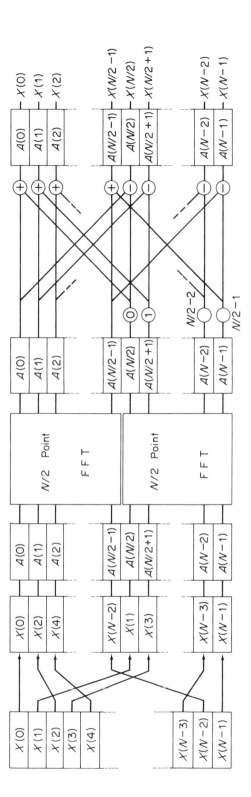

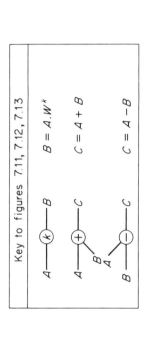

Key to figures 7.11, 7.12, 7.13

$A \longrightarrow \!\!\boxed{k}\!\! \longrightarrow B \qquad B = A.W^k$

$A \searrow \!\!\boxed{+}\!\!\longrightarrow C \qquad C = A + B$
$B \nearrow$

$A \searrow \!\!\boxed{-}\!\!\longrightarrow C \qquad C = A - B$
$B \nearrow$

**Figure 7.11**   The first stage in forming the fast Fourier transform

Hence, for the $N/2$-point transforms we shall require four equations:

$$A(n) \leftarrow \tfrac{1}{2}\{A(n) + A(n + N/4) . W^{2n}\}$$

$$A(n + N/4) \leftarrow \tfrac{1}{2}\{A(n) - A(n + N/4) . W^{2n}\}$$

$$A(n + N/2) \leftarrow \tfrac{1}{2}\{A(n + N/2) + A(n + 3N/4) . W^{2n}\}$$

$$A(n + 3N/4) \leftarrow \tfrac{1}{2}\{A(n + N/2) - A(n + 3N/4)W^{2n}\}$$

all for the other shorter range of $n$:

$$n = 0, 1, \ldots, \left(\frac{N}{4} - 1\right).$$

Provided that $N$ is an integral part of two:

$$N = 2^r.$$

This procedure of repeatedly decimating the input data and rewriting the operational equations may be performed $r$ times until only two-point transforms remain. The entire transform operation then consists of a succession of weightings and additions of the data stored in the array $A(n)$. As an example, the complete computation flowgraph is illustrated in Figure 7.12 for the particular case where $N = 8$.

As Figure 7.12 demonstrates, the input data $x(k)$ is applied to the array $A(n)$ in shuffled, not natural order. This is the result of the repeated decimation process which is performed on the data as the transforms are broken down. The shuffling operation is neatly described in the following way. If an input datum subscript is written in binary form, for example, $x(6)$ becomes $x(110)$, then the location of the shuffled datum is obtained by reversing the order of the binary digits. Thus, in the example quoted above, $x(110)$ becomes $x(011)$, which translates back into decimal to give the location as 3. Thus we place $x(6)$ in the third location and $x(3)$ in the sixth. The entire shuffle is referred to as 'bit reversal'.

Another interesting feature of the flow diagram is the 'inplace' nature of the computation. As we have seen, only as much storage is needed as is required to hold the initial $N$ data points. As the computation proceeds, the initial data are destroyed and replaced first with intermediate results and finally with other naturally ordered transform values.

The transform method we have discussed is referred to as 'decimation in time'. This is simply because it is the input time sequence which is broken down into subsequences. Another quite distinct implementation can be derived by breaking down the transform sequence. This method is referred to as 'decimation in frequency' and yields bit reversed values of the transform from naturally ordered data. It is interesting to note that, although the procedure described here obviously requires that $N$ is an integral power of two, it can be modified to work for other values of $N$. However, the computation can be shown to become rather less efficient, particularly if the numbers into which $N$ can be factorized are not small. In fact, if $N$ is a prime number, no fast algorithm exists, and the simple implementation of the DFT illustrated in Figure 7.9 must be employed.

111

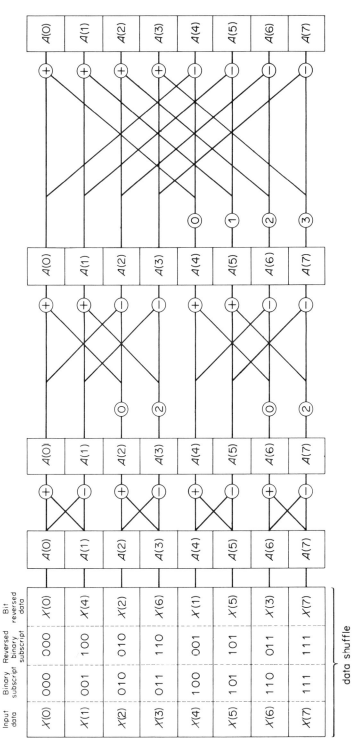

**Figure 7.12** Flowgraph for the fast Fourier transform

We are now in a position to estimate the computational efficiency of the fast transform algorithm and then deduce and implement a set of recursive equations describing the FFT.

### 7.6 Computational efficiency in the FFT

The obvious and simple implementation of the DFT referred to earlier, and depicted in Figure 7.9, requires $N$ complex multiplications to evaluate each of the $N$ values of $A(n)$. Since multiplication is the most time consuming of the basic arithmetic and manipulative operations available on a digital computer, it follows that the time required to calculate the DFT in this manner will be closely proportional to $N^2$.

In contrast, if we examine the flowchart in Figure 7.12, we find that the FFT requires $N/2 . \log_2 N$ complex multiplications. This is because we can obtain the final array or any intermediate array $A(n)$ from its predecessor, with $(N/2)$ weightings, each of which involves a single complex multiplication. The weightings occur $r$ times, and

$$r = \log_2 N.$$

For large values of $N$, the FFT becomes very economical in terms of computation time. For example, if $N = 1024$, the FFT is over one hundred times faster than a straightforward DFT.

### 7.7 A set of recursive equations for the FFT

We may conveniently identify the contents of the array $A(n)$ at successive points throughout the flowchart by using a subscript notation. Thus the leftmost array in the flowchart of Figure 7.13 becomes $A_1(n)$, the rightmost $A_{r+1}(n)$. Let the general array be $A_i(n)$, with $i$ covering the range from 1 to $r + 1$ and corresponding to $r$ successive sets of weighting operations on the input data. As an example, the final weighting operation is then:

$$A_{r+1}(n) \leftarrow \tfrac{1}{2}\left\{A_r(n) + A_r\left(n + \frac{N}{2}\right)W^n\right\}$$

$$A_{r+1}\left(n + \frac{N}{2}\right) \leftarrow \tfrac{1}{2}\left\{A_r(n) - A_r\left(n + \frac{N}{2}\right)W^n\right\}.$$

When we generalize to give the $A_{i+1}(n)$ in terms of the $A_i(n)$ we must bear in mind several important points. As we move to the left across the flowchart in Figure 7.13:

(a) the separation between the pairs of locations which are manipulated during each stage of weighting will halve. As a consequence:
(b) the number of operational statements will double.

113

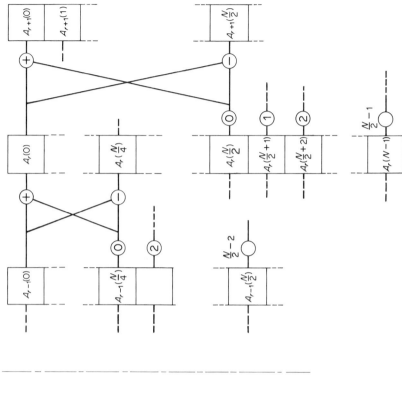

**Figure 7.13**   Illustrating the recursive algorithm for the FFT

(c) the range of the independent variable must halve. Finally:
(d) the argument of the weighting function must double.

To take the first of these requirements into account we manipulate locations

$$A_i(n) \quad \text{and} \quad A_i(n + 2^{i-1}).$$

The second and third conditions are the most difficult to contend with. These we may satisfy by writing

$$n = (m - 1) + (l - 1) \cdot 2^i.$$

We have, then, $l$ pairs of operational statements, with

$$l = 1, 2, \ldots, 2^{r-i}$$

and an independent variable $m$ of range

$$m = 1, 2, \ldots, 2^{i-1}.$$

Note that, irrespective of the value of $i$, the variable $n$ always takes on all its $N$ values as $l$ and $m$ cycle over their entire ranges. However, the order of selection of the values *does* depend on $i$.

Finally, we contend with condition (d) by expressing the weighting function as: $W^{n.2^{r-i}}$. We may simplify this function by writing $n$ in terms of $m$ and $l$, noting that $W^{(l-1)N} = 1$; all integer and we finally obtain $W^{(m-1)2^{r-i}}$. We are now in a position to construct the final recursive equations, and these are:

$$A_{i+1}[(m - 1) + (l - 1) \cdot 2^i] = \tfrac{1}{2}\{A_i[(m - 1) + (l - 1) \cdot 2^i]$$
$$+ A_i[(m - 1) + (l - 1) \cdot 2^i + 2^{i-1}] \cdot W^{(m-1) \cdot 2^{r-i}}\}$$

$$A_{i+1}[(m - 1) + (l - 1) \cdot 2^i + 2^{i-1}] = \tfrac{1}{2}\{A_i[(m - 1) + (l - 1) \cdot 2^i]$$
$$- A_i[(m - 1) + (l - 1) \cdot 2^i + 2^{i-1}] \cdot W^{(m-1).2^{r-i}}\}$$

for:

$$i = 1, 2, \ldots, r$$

$$m = 1, 2, \ldots, 2^{i-1}$$

$$l = 1, 2, \ldots, 2^{r-i}. \tag{7.22}$$

## 7.8 Programming the algorithm

Now that we have obtained the recursive equations describing the FFT, the actual programming becomes fairly straightforward. Notice that the computation requires three 'loops' covering the ranges of $i$, $l$ and $m$, respectively. The least rapidly changing of these must be $i$, since both $m$ and $l$ depend on its value to determine the limit of their ranges. A flowchart describing the implementation of the recursive equations is shown in Figure 7.14. From this, it is an easy task to derive a FORTRAN program or subprogram which performs the

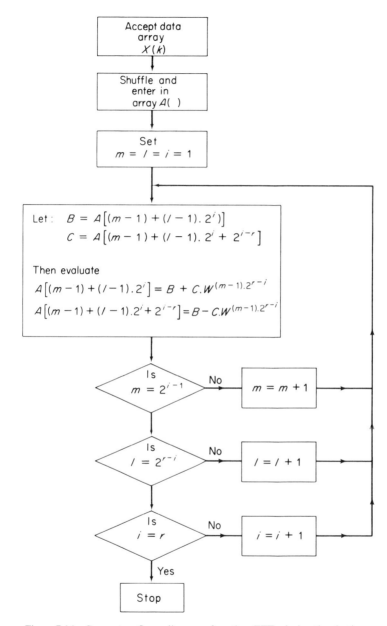

**Figure 7.14** Computer flow diagram for the FFT decimation-in-time algorithm

FFT. Figure 7.15a illustrates subroutine 'EASY', a fairly typical simple implementation. As the 'comment' cards indicate, the first few statements perform bit reversal. The remainder is concerned with performing the decimation-in-time FFT.

```
          SUBROUTINE EASY(A,R)
          INTEGER R
          REAL A(2,1)
          PI = 3.1415926
          N = 2**R
          LIM1 = N - 1
          LIM2 = N/2
     C    INPLACE SHUFFLE OF DATA BEGINS
          J = 1
          DO 3 I = 1,LIM1
          IF(I.GE.J) GO TO 1
          A1 = A(1,J)
          A2 = A(2,J)
          A(1,J) = A(1,I)
          A(2,J) = A(2,I)
          A(1,I) = A1
          A(2,I) = A2
        1 L = LIM2
        2 IF(L.GE.J) GO TO 3
          J = J - L
          L = L/2
          GO TO 2
        3 J = J + L
     C    SHUFFLE COMPLETE
     C    TRANSFORM BY DECIMATION IN TIME BEGINS
          DO 4 I = 1,R
          LIM1 = 2**(I - 1)
          LIM2 = 2**(R - I)
          DO 4 L = 1,LIM2
          DO 4 M = 1,LIM1
          LIM3 = (M - 1) + (L - 1)*2*LIM1 + 1
          B1 = A(1,LIM3)
          B2 = A(2,LIM3)
          C1 = A(1,LIM3 + LIM1)
          C2 = A(2,LIM3 + LIM1)
          ARG = 2.0*PI*FLOAT((M - 1)*LIM2)/FLOAT(N)
          A1 = C1*COS(ARG) + C2*SIN(ARG)
          A2 =-C1*SIN(ARG) + C2*COS(ARG)
          A(1,LIM3) = B1 + A1
          A(2,LIM3) = B2 + A2
          A(1,LIM3 + LIM1) = B1 - A1
          A(2,LIM3 + LIM1) = B2 - A2
        4 CONTINUE
          END
```

**Figure 7.15**(a)  FORTRAN programme to permit computation
of DFT kernel as defined by equation (7.16)

A further improvement in computational efficiency, illustrated in the version of 'EASY' provided in Figure (7.15b), is obtained when we calculate the sine and cosine values of the weighting function by a recursive method. This saves calling upon the library functions COS and SIN, both of which are time-consuming operations. The recursion formulae are easily derived by a trigonometric expansion of general $(k + 1)$th terms $\cos[(k + 1)\theta]$ and $\sin[(k + 1)\theta]$ where:

$$\theta = 2\pi/N(2^{r-i}); \qquad i = 1, 2, \ldots, r.$$

As it stands, EASY performs the calculation defined by equation (7.16). To perform the DFT operaton, as defined by equations (7.17) and (7.18) we follow

```
      SUBROUTINE EASY(A,R)
      INTEGER R
      REAL A(2,1)
      PI = 3.1415926
      N = 2**R
      LIM1 = N - 1
      LIM2 = N/2
C     INPLACE SHUFFLE OF DATA BEGINS
      J = 1
      DO 3 I = 1,LIM1
      IF(I.GE.J) GO TO 1
      A1 = A(1,J)
      A2 = A(2,J)
      A(1,J) = A(1,I)
      A(2,J) = A(2,I)
      A(1,I) = A1
      A(2,I) = A2
    1 L = LIM2
    2 IF(L.GE.J) GO TO 3
      J = J - L
      L = L/2
      GO TO 2
    3 J = J + L
C     SHUFFLE COMPLETE
C     TRANSFORM BY DECIMATION IN TIME BEGINS
      DO 4 I = 1,R
      LIM1 = 2**(I - 1)
      LIM2 = 2**(R - 1)
      ARG = 2.0*PI*LIM2/FLOAT(N)
      CS = 1.0
      SI = 0.0
      CSTEP = COS(ARG)
      SSTEP = SIN(ARG)
      DO 4 M = 1,LIM1
      DO 5 L = 1,LIM2
      LIM3 = (M - 1) + (L - 1)*2*LIM1 + 1
      B1 = A(1,LIM3)
      B2 = A(2,LIM3)
      C1 = A(1,LIM3 + LIM1)
      C2 = A(2,LIM3 + LIM1)
      A1 = C1*CS + C2*SI
      A2 =-C1*SI + C2*CS
      A(1,LIM3) = B1 + A1
      A(2,LIM3) = B2 + A2
      A(1,LIM3 + LIM1) = B1 - A1
      A(2,LIM3 + LIM1) = B2 - A2
    5 CONTINUE
      CS1 = CS*CSTEP - SI*SSTEP
      SI1 = SI*CSTEP + CS*SSTEP
      CS = CS1
      SI = SI1
    4 CONTINUE
      END
```

**Figure 7.15**(b)  Improved DFT programme employing recursive
generation of weights

the call to EASY by a call to subroutine SCALE, Figure 7.16. Finally to perform the IDFT, as defined by equations (7.19) and (7.20) we precede the call to EASY by a call to SORT, Figure 7.17.

```
SUBROUTINE SCALE(X,R)
DIMENSION X(2,1)
INTEGER R
N = 2**R
DO 1 I = 1,N
X(1,1) = X(1,I)/FLOAT(N)
X(2,1) = X(2,I)/FLOAT(N)
1 CONTINUE
RETURN
END
```

**Figure 7.16** Programme to scale the result obtained from 'EASY' by a factor $(1/N)$

```
SUBROUTINE SORT(X,R)
INTEGER R
REAL X(2,1)
N = 2**R
LIM1 = N/2
DO 1 I = 2,LIM1
X1 = X(1,I)
X2 = X(2,I)
LIM2 = N - I + 2
X(1,I) = X(1,LIM2)
X(2,I) = X(2,LIM2)
X(1,LIM2) = X1
X(2,LIM2) = X2
1 CONTINUE
RETURN
END
```

**Figure 7.17** Programme to perform the data reversal required to obtain an IDFT by means of the DFT programme 'EASY'

## 7.9 Computing the DFT of real sequences

In the foregoing sections of this chapter we have examined the DFT and its implementation by means of the FFT. Both the DFT and the FFT process data $x(k)$ to produce a spectrum $X(n)$. Although $x(k)$ is, in its most general form, complex, most engineering problems only involve real variables:

$$\text{Im}\,[x(k)] = 0.$$

Such functions possess a spectrum which exhibits complex conjugate symmetry:

$$X(n) = X^*(N - n) \qquad n = 1, 2, \ldots, (N/2 - 1).$$

Consequently, if the FFT were used directly to evaluate the spectrum of a real $N$-point function, half the total storage requirement of $2N$ locations could be regarded as redundant. Furthermore, computing time would be wasted in evaluating the negative frequency half of the spectrum. We shall now examine two methods of processing real data which eliminate these sources of inefficiency. Both make use of a fundamental symmetry property, that any asymmetric function may be formed as the sum of two functions, symmetric about some suitable axis, one possessing even and the other odd symmetry. For our future convenience we may state this property thus:

$$X(n) = X_{\text{even}}(n) + X_{\text{odd}}(n); \qquad n = 0, 1, \ldots, N - 1$$

where:

$$X_{even}(n) = X_{even}(N - n) = \frac{X(n) + X(N - n)}{2}$$

and:

$$X_{odd}(n) = -X_{odd}(N - n) = \frac{X(n) - X(N - n)}{3}$$

both for $n = 1, 2, \ldots, (N/2 - 1)$. We also define:

$$X_{even}(0) = X(0)$$

$$X_{even}(N/2) = X(N/2)$$

$$X_{odd}(0) = 0$$

$$X_{odd}(N/2) = 0.$$

The first of the two methods we shall discuss enables us to compute the positive frequency portion of the spectra of two equal length $N$-point real functions simultaneously, using one $N$-point complex FFT. It is also the first stage of processing required for the second method, by means of which we may compute the positive frequency half of the spectrum of one real $N$-point function, using an $(N/2)$-point complex FFT.

Figure 7.18(a) shows two real, $N$-point functions $y(k)$ and $z(k)$. We multiply $z(k)$ by the operator j to produce a purely imaginary function. The spectra $Y(n)$ and $Z(n)$ both exhibit complex conjugate symmetry about the point $n = N/2$, Figure 7.18(b). However, the symmetries are such that the transform, $X(n)$, of the sum: $x(k) = y(k) + jz(k)$ is asymmetric, Figure 7.18(c). The original spectra, $Y(n)$ and $Z(n)$, may be reconstructed from $X(n)$, over the positive frequency part of their range, by a simple data manipulation obtained by applying the symmetry property quoted above:

$$Re\,[Y(n)] = Re\,[X(n) + X(N - n)]/2$$

$$Im\,[Y(n)] = Im\,[X(n) - X(N - n)]/2$$

$$Re\,[Z(n)] = Im\,[Z'(n)]$$

$$= Im\,[X(n) + X(N - n)]/2$$

$$Im\,[Z(n)] = -Re\,[Z'(n)]$$

$$= -Re\,[X(n) - X(N - n)]/2 \qquad (7.23)$$

all for $n = 1, 2, \ldots, (N/2 - 1)$ and with

$$Re\,[Y(0)] = Re\,[X(0)]$$

$$Re\,[Z(0)] = Im\,[X(0)]$$

$$Re\,[Y(N/2)] = Re\,[X(N/2)]$$

$$Re\,[Z(N/2)] = Im\,[X(N/2)]. \qquad (7.24)$$

120

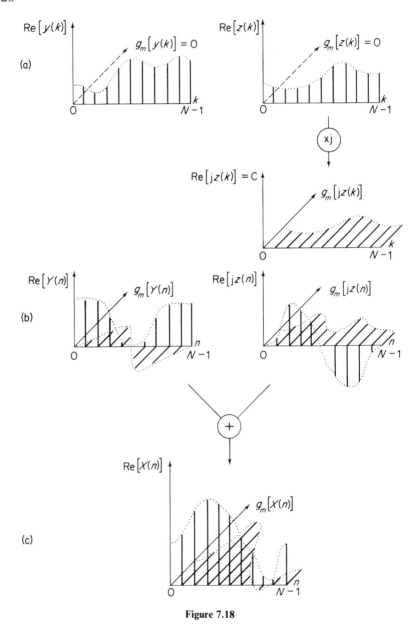

**Figure 7.18**

Since, from a computational standpoint, we only allow $N/2$ storage locations for each of the real and imaginary parts of the two functions $Y(n)$ and $Z(n)$, the midpoint values $Y(N/2)$ and $Z(N/2)$ are most conveniently held thus:

$$\text{Im}\,[Y(0)] = \text{Re}\,[Y(N/2)]$$

$$\text{Im}\,[Z(0)] = \text{Re}\,[Z(N/2)].$$

```
SUBROUTINE RFT1(A,R)
INTEGER R
DIMENSION A(2,1)
CALL SCALE(A,R)
CALL EASY(A,R)
N = 2**R
LIM1 = N/2
LIM2 = LIM1 + 1
LIM3 = N - 1
A1 = A(2,1)
A(2,1) = A(1,LIM2)
A2 = A(2,LIM2)
DO 1 I = 2,LIM1
A3 = A(1,I)
A4 = A(2,I)
A5 = A(1,N - 1 + 2)
A6 = A(2,N - 1 + 2)
A(1,I) = (A5 + A3)/2.0
A(2,I) = (A4 - A6)/2.0
A(1,N - 1 + 2) = (A4 + A6)/2.0
A(2,N - 1 + 2) = (A5 - A3)/2.0
1 CONTINUE
DO 2 I = LIM2,LIM3
A(1,I) = A(1,I + 1)
A(2,I) = A(2,I + 1)
2 CONTINUE
A(1,N) = A1
A(2,N) = A2
RETURN
END
```

**Figure 7.19** Programme to permit the computation of the DFT's of two real $N$-point functions

(Recall that, for a real function the imaginary part of the midpoint spectral value is zero, and need not, therefore, be computed). Figure 7.19 lists a routine, RFT1, which may be used to compute the DFT of two functions in the manner described above. One function is inserted as the real part of the input array $A(1, I)$, the other as the imaginary part, $A(2, I)$:

| I | | 1 | 2 | ... | N |
|---|---|---|---|---|---|
| Real part | $A(1, I)$ | $y(0)$ | $y(1)$ | ... | $y(N - 1)$ |
| Imag. part | $A(2, I)$ | $z(0)$ | $z(1)$ | ... | $z(N - 1)$ |

After the in-place transformation and manipulation of the storage locations containing the results, the spectra of $y(k)$ and $z(k)$ are held thus:

| I | | 1 | 2 | | $N - 1$ | $N$ |
|---|---|---|---|---|---|---|
| Real part | $A(1, I)$ | Re$[Y(0)]$ | Re$[Y(1)]$ | ... | Re$[Z(1)]$ | Re$[Z(0)]$ |
| Imag. part | $A(2, I)$ | Re$[Y(N/2)]$ | Im$[Y(1)]$ | ... | Im$[Z(1)]$ | Re$[Z(N/2)]$ |

The inversion algorithm is easily obtained from equations (7.23) by solving to give $X(n)$ and $X(N - n)$ in terms of $Y(n)$ and $Z(n)$:

$$\text{Re}\,[X(n)] = \text{Re}\,[Y(n)] - \text{Im}\,[Z(n)]$$

$$\text{Im}\,[X(n)] = \text{Re}\,[Z(n)] + \text{Im}\,[Y(n)]$$

$$\text{Re}\,[X(N - n)] = \text{Re}\,[Y(n)] + \text{Im}\,[Z(n)]$$

$$\text{Im}\,[X(N - n)] = \text{Re}\,[Z(n)] - \text{Im}\,[Y(n)]$$

all for $n = 1, 2, \ldots, (N/2 - 1)$.

The values corresponding to $n = 0$ and $N/2$ are given by equations (7.24).

Figure 7.20 illustrates the subroutine, RFT2, employed to perform the inversion operation on the transforms of two real sequences.

```
SUBROUTINE RFT2(A,R)
INTEGER R
DIMENSION A(2,1)
N = 2**R
LIM1 = N/2
A1 = A(1,N)
A2 = A(2,N)
DO 1 I = 1,LIM1
A(1,N-I+1) = A(1,N-I)
A(2,N-I+1) = A(2,N-I)
1 CONTINUE
A3 = A(2,1)
A(2,1) = A1
A(2,LIM1 + 1) = A2
A(1,LIM1 + 1) = A3
DO 2 I = 2,LIM1
A5 = A(1,1) - A(2,N-I+2)
A6 = A(2,I) - A(1,N-I+2)
A7 = A(1,1) + A(2,N-I+2)
A8 = A(2,1) + A(1,N-I+2)
A(1,I) = A5
A(2,N-I+2) =-A6
A(1,N-I+2) = A7
A(2,I) = A8
2 CONTINUE
CALL SORT(A,R)
CALL EASY(A,R)
RETURN
END
```

**Figure 7.20** Programme to obtain two real functions from their ordered and stored DFT's (for details of the ordering of the DFT's within the array A see text. This programme complements 'RFT 1')

The method of computing the DFT of one real $N$-point function closely parallels the initial stage of development of the FFT itself. Recall that the spectrum of a function may be obtained from the spectra of its decimation components. In this case, we accept a real function

$$x_1(k) \qquad k = 0, 1, \ldots, N - 1$$

and decimate to form two real $N/2$-point functions:

$$y(k) = x_1(2k)$$
$$z(k) = x_1(2k + 1)$$
$$k = 0, 1, \ldots, (N/2 - 1).$$

The use, once again, of the functions $y$ and $z$ is intended to establish a direct connection with the results obtained above. The only difference lies in the length of those two functions, each of which now consists of $N/2$ real values, instead of $N$.

By direct analogy with equation (7.19) we see that:

$$X_1(n) = \tfrac{1}{2}[Y(n) + W^n . Z(n)]; \qquad n = 0, 1, \ldots, (N/2 - 1).$$

Since $Y(n)$ and $Z(n)$ consist implicitly of $N/4$ complex (i.e. $N/2$ real) values it is computationally convenient for us to break this equation into two operational statements:

$$X_1(n) = \tfrac{1}{2}[Y(n) + W^n . Z(n)]$$
$$X_1(N/2 - n) = \tfrac{1}{2}[Y(N/2 - n) + W^{(N/2-n)}Z(N/2 - n)]$$

(7.25)

both for $n = 0, 1, \ldots, (N/4 - 1)$. The second of these statements may be written in terms of the available $N/4$ complex values of $Y(n)$ and $Z(n)$ because

$$Y(N/2 - n) = Y^*(n)$$

and

$$Z(N/2 - n) = Z^*(n).$$

Thus, for the same range of $n$:

$$X_1(N/2 - n) = \tfrac{1}{2}[Y^*(n) - W^{-n}Z^*(n)]. \tag{7.26}$$

We may compute $Y(n)$ and $Z(n)$ simultaneously by calling upon subroutine RFT1 and suitably choosing the value of $R$ (which defines the lengths of $Y(n)$ and $Z(n)$ as $2^R$) in its argument list. This call, followed by the data manipulation defined by equations (7.23) and (7.24) yields the spectrum $X_1(n)$. The entire computation is performed by subroutine RFT3, Figure 7.21. Here, the input array $A$ has as its real part of the sequence $y(k)$ and, as its imaginary part, the sequence $z(k)$:

| I | | 1 | 2 | $\ldots$ | $N/2$ |
|---|---|---|---|---|---|
| Real part | $A(1, I)$ | $Y(0)$ | $Y(1)$ | $\cdots$ | $Y(N/2 - 1)$ |
| Imag. part | $A(2, I)$ | $Z(0)$ | $Z(1)$ | $\ldots$ | $Z(N/2 - 1)$ |

This corresponds to inserting the $N$-point real function thus:

| I | | 1 | 2 | ... | $N/2$ |
|---|---|---|---|-----|-------|
| Real part | $A(1, I)$ | $x(0)$ | $x(2)$ | ... | $x(N - 2)$ |
| Imag. part | $A(2, I)$ | $x(1)$ | $x(3)$ | ... | $x(N - 1)$ |

```
SUBROUTINE RFT3(A,R)
INTEGER R
DIMENSION A(2,1)
PI = 3.14159
CALL RFT1(A,R)
N = 2**R
LIM1 = N/2
LIM2 = LIM1 - 1
A1 = A(2,1)/2.0
A2 = A(2,N)/2.0
A3 = A(1,1)/2.0
A4 = A(1,N)/2.0
DO 1 I = 2,LIM1
A5 = A(1,I)/2.0
A6 = A(2,I)/2.0
A7 = A(1,N - I + 1)/2.0
A8 = A(2,N - I + 1)/2.0
ARG = PI*FLOAT(I - 1)/FLOAT(N)
A(1,I) = A5 + A7*COS(ARG) + A8*SIN(ARG)
A(2,I) = A6 - A7*SIN(ARG) + A8*COS(ARG)
A(1,N - I + 1) = A5 - A7*COS(ARG) - A8*SIN(ARG)
A(2,N - I + 1) =-A6 - A7*SIN(ARG) + A8*COS(ARG)
1 CONTINUE
DO 2 I = 1,LIM2
A(1,N - I + 1) = A(1,N + I)
A(2,N - I + 1) = A(2,N - 1)
2 CONTINUE
A(1,LIM1 + 1) = A1
A(2,LIM1 + 1) =-A2
A(1,1) = A3 + A4
A(2,1) = A3 - A4
RETURN
END
```

**Figure 7.21**   Programme to compute the DFT of one real $N$-point
function

It is worth noting that such an insertion of data can be very easily achieved when programming in FORTRAN IV by making use of the EQUIVALENCE declaration. Reference to the listings of subroutines FILTR (Figure 7.25) and WKIN (Figure 7.28) illustrate this point.

The spectrum of $x(k)$ is presented by RFT3 in array $A$, in the form:

| I | | 1 | 2 | ... | $N/2$ |
|---|---|---|---|-----|-------|
| Real part | $A(1, I)$ | $\mathrm{Re}\,[X(0)]$ | $\mathrm{Re}\,[X(1)]$ | ... | $\mathrm{Re}\,[X(N/2 - 1)]$ |
| Imag. part | $A(2, I)$ | $\mathrm{Im}\,[X(0)]$ | $\mathrm{Im}\,[X(1)]$ | ... | $\mathrm{Im}\,[X(N/2 - 1)]$ |

Inversion, to obtain the original $N$-point real function may be readily achieved by solving equation (7.25) and (7.26) to yield $Y(n)$ and $Z(n)$ in terms of $X_1(n)$ and $X_1(N/2 - n)$:

$$Y(n) = X_1(n) + X_1^*(N/2 - n)$$

$$Z(n) = (X_1(n) - X_1^*(N/2 - n)) . W^{-n}.$$

These equations are implemented by means of subroutine RFT4, shown in Figure 7.22.

```
SUBROUTINE RFT4(A,R)
INTEGER R
DIMENSION A(2,1)
PI = 3.14159
N = 2**R
LIM1 = N/2
LIM2 = LIM1 + 1
LIM3 = N - 1
A1 = A(1,LIM2)*2.0
A2 = A(2,LIM2)*2.0
A3 = A(1,1)
A4 = A(2,1)
DO 1 I = LIM2,LIM3
A(1,1) = A(1,I+1)
A(2,I) = A(2,I+1)
1 CONTINUE
DO 2 I = 2,LIM1
ARG = PI*FLOAT(I-1)/FLOAT(N)
A5 = A(1,1)
A6 = A(2,1)
A7 = A(1,N-I+1)
A8 = A(2,N-I+1)
A(1,1) = A5 + A7
A(2,I) = A6 - A8
A(1,N-I+1) = (A5 - A7)*COS(ARG) - (A6 + A8)*SIN(ARG)
A(2,N-I+1) = (A5 - A7)*SIN(ARG) + (A6 + A8)*COS(ARG)
2 CONTINUE
A(1,1) = A3 + A4
A(1,N) = A3 - A4
A(2,1) = A1
A(2,N) = -A2
CALL RFT2(A,R)
RETURN
END
```

**Figure 7.22**   Programme to compute the IDFT of the positive-frequency half of the spectrum of a real $N$-point function

## 7.10   The convolution of long sequences

We saw, in Section 7.2.9, that it is possible to convolve two equal length periodic sequences $x(k)$ and $h(k)$ by multiplying their DFT's $X(n)$ and $H(n)$ and applying the IDFT to the result. The operation

$$y(m) = \sum_{n=0}^{N-1} [X(n)H(n)]W^{nm}; \qquad m = 0, 1, \ldots, N - 1 \qquad (7.27)$$

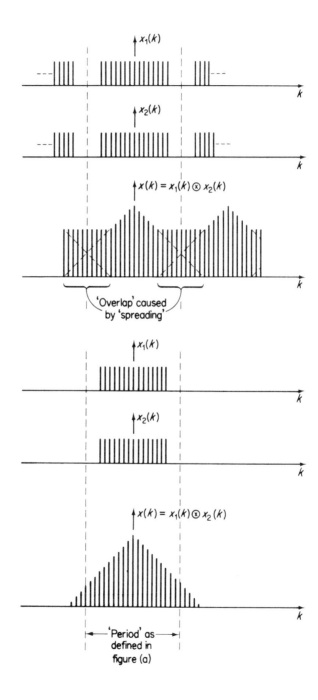

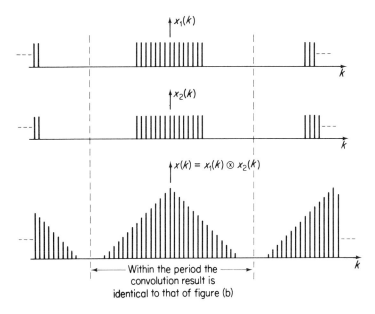

**Figure 7.23**(a) Correct convolution of periodic rectangular function. (b) Correct convolution of aperiodic rectangular function. (c) Effect of increasing the period in a periodic convolution

corresponds exactly to the true convolution of $x(k)$ and $h(k)$

$$y(m) = \sum_{i=0}^{N-1} x(i)h(m-i); \quad m = 0, 1, \ldots, N-1. \tag{7.28}$$

Normally, we are interested in employing the principle of convolution to achieve a filtering action. When this is the case, both our data, $x(k)$, and the filter impulse response, $h(k)$, are aperiodic. An example of such a convolution is that which defines the $L$-tap, or non-recursive filter which has been examined in detail in Section 6.

$$y(m) = \sum_{i=0}^{L-1} x(i) . h(m-i); \quad \text{all } m. \tag{7.29}$$

To emphasize the fact that in general we can only rely on equations (7.27) or (7.28) to produce a correct convolution when $x(k)$ and $h(k)$ are both periodic, consider the example given in Figure 7.23.

First we convolve two periodic rectangular waveforms, obtaining the correct periodic convolution, Figure 7.23(a). Next, we perform the convolution of the two related aperiodic rectangular pulses, Figure 7.23(b). The result of this is distinctly different. The 'spreading' inherent in the convolution operation causes an 'overlap' when the periodic convolution is performed. If we require a

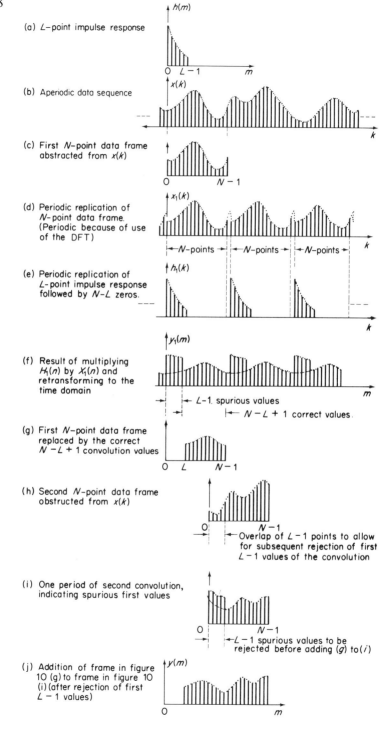

(a) $L$-point impulse response

(b) Aperiodic data sequence

(c) First $N$-point data frame abstracted from $x(k)$

(d) Periodic replication of $N$-point data frame. (Periodic because of use of the DFT)

(e) Periodic replication of $L$-point impulse response followed by $N-L$ zeros.

(f) Result of multiplying $H_1(n)$ by $X_1(n)$ and retransforming to the time domain

(g) First $N$-point data frame replaced by the correct $N-L+1$ convolution values

(h) Second $N$-point data frame obstructed from $x(k)$

(i) One period of second convolution, indicating spurious first values

(j) Addition of frame in figure 10 (g) to frame in figure 10 (i) (after rejection of first $L-1$ values)

numerical evaluation of the convolution of aperiodic functions, then, care must be taken to ensure that this overlap does not occur. In our example this could be achieved by increasing the period of the functions, Figure 7.23(c), and employing a larger DFT.

The use of the DFT in equation (7.27) implies that $x(k)$ and $h(k)$ are assumed to be periodic. Since periodic functions always produce a periodic convolution result, discrete convolutions employing the DFT are often referred to as 'circular'.

When the convolution is intended to perform a filtering operation, a second difficulty is often encountered. Normally, we filter long sequences of data against a relatively short impulse response. This results in a situation which can be extremely costly of computer storage. We can use the DFT to perform such convolutions by a suitable sectioning of the long data sequence.

Consider the $L$-point impulse response $h(m)$ shown in Figure 7.24(a). This is to be convolved with $x(k)$, a long data sequence, Figure 7.24(b). To achieve this, we abstract $N$ data points from $x(k)$ forming a 'frame' which may be used as input data to an $N$-point DFT, Figure 7.24(c). For a reason which should become apparent as the analysis proceeds, we place a restriction on $N$ which requires that

$$N > L.$$

Because we employ the DFT, the 'frame' must be regarded as one period of a periodic function, $x_1(t)$. This function is shown in Figure 7.24(a). Next we evaluate, again by means of a DFT, the transform of an $N$-point function formed by appending $N-L$ zeros to our $L$-point impulse response. This implies that a periodic repetition of $h(k)$ has been established, $h_1(k)$ in Figure 7.24(e), for convolution with $x_1(k)$ by frequency domain multiplication:

$$Y_1(n) = X_1(n) . H_1(n)$$

where

$$X_1(n) = \frac{1}{N} \sum_{k=0}^{N-1} x_1(k)W^{-nk} \qquad n = 0, 1, \ldots, N-1$$

and

$$H_1(n) = \frac{1}{N} \sum_{k=0}^{N-1} h_1(k)W^{-nk}.$$

We obtain $y_1(m)$ by applying the IDFT:

$$y_1(m) = \sum_{n=0}^{N-1} [X_1(n)H_1(n)]W^{nm}; \qquad m = 0, 1, \ldots, N - 1.$$

If we insert the two equations given above, which yield $X_1(n)$ and $H_1(n)$, and apply the orthogonality relation, Section 7.2(e), we can derive the result:

$$y_1(m) = \sum_{i=0}^{m} h_1(i)x_1(m - i) + \sum_{i=m+1}^{N-1} h_1(i)x_1(N + m - i). \qquad (7.30)$$

Comparison of this equation with the defining equation for a non-recursive filter, equation (7.29), indicates similarities, but clearly the two are not identical. Recall that we have specified that $N$ must be greater than $L$, and consider the case for which $m = L - 1$:

$$y_1(L - 1) = \sum_{i=0}^{L-1} h_1(i)x_1(L - 1 - i) + \sum_{i=L}^{N-1} h_1(i)x_1(N + L - 1 + i).$$

Here,

$$h_1(L) = h_1(L + 1) = \cdots = h_1(N - 1) = 0$$

so that the second term may be deleted and

$$y_1(L - 1) = \sum_{i=0}^{L-1} h_1(i)x_1(L - 1 - i).$$

This has precisely the correct form, and gives the required $(L - 1)$th output sample from an $L$-tap filter. Again, the terms $y_1(L), y_1(L + 1), \ldots, y_1(N - 1)$ will also be correct, for the same reason.

We must now consider the first $(L - 1)$ output samples. These all contain spurious 'overlap' terms caused by the second summation in equation (7.30). They are therefore incorrect and must be rejected as invalid data. This effect is illustrated in Figures 7.24(f) and 7.24(g).

The next operation consists of extracting a second $N$-point data frame, so chosen that its first $(L - 1)$ values are identical with the last $(L - 1)$ points taken from $x(k)$ to form the first data frame, Figure 7.24(h). This is because the next $N$-point convolution will again yield $(L - 1)$ spurious initial values which will have to be discarded.

We return the valid part of the result of the first convolution to store, Figure 7.24(i). Since we have now extracted at least $N$ points from $x(k)$ which need never be used again, the filtered data so far accumulated as a result of the first convolution may be placed in the same storage locations as contained the first prefiltered data frame. Thus the whole filtering operation may be performed 'in-place', and, if this policy is adopted, will require only the storage necessary to hold the initial data. A relatively small amount of 'scratch' storage (the exact quantity of which will depend on the detailed organization of the computation) will also be needed to perform the actual convolution operation and transforms.

The abuttal of the valid parts of the first and second convolution results is shown in Figure 7.24(j) and illustrates the manner in which the complete convolution may be assembled from the much smaller $N$-point convolutions.

The technique which has just been described is known as the 'select-save' method[8], for reasons which should now be apparent. The 'overlap-add' method[8], is another means of achieving the same end. Although it differs in detail from the select-save method, it is based on the same principle and requires the same computing time and storage.

It will be appreciated that if $N$ is made large relative to $L$, which we presume to be fixed (determined by the required number of filter taps), then large segments of the filter output will be calculated each time the convolution is performed. Consequently, relatively few convolutions will be needed to filter a given length input sequence. On the other hand if $N$ is small, not much larger than $L$, small, quick convolutions will be needed, but there will be many more to do in filtering $x$. In fact, optimum values of $N$ for given values of $L$ have been calculated[8] and these are given in Table 7.1.

Table 7.1

| $L$ | $N$ |
| --- | --- |
| $<11$ | 32 |
| 11– 17 | 64 |
| 18– 29 | 128 |
| 30– 52 | 256 |
| 53– 94 | 512 |
| 95– 171 | 1024 |
| 172– 310 | 2048 |
| 311– 575 | 4096 |
| 576–1050 | 8192 |
| 1051–2000 | 16,384 |
| 2001–3800 | 32,768 |
| 3801–7400 | 65,536 |
| $>7400$ | 131,072 |

Finally Figure 7.25 lists a subroutine, FILTR, which performs the 'select-save' convolution. The subroutine accepts an input data array $X$ of specified length and convolves it with an externally defined impulse response AH of length:

$$N = 2^R.$$

The duration, $L$, of the non-zero part of the impulse response array must be selected in accordance with the bounds set in Table 7.1. Both the impulse response and the input data are real, so that FILTR provides an efficient method of processing most engineering functions which require numerical filtering.

We are now in a position to compare the computation speed of convolutions performed using the 'select-save' or 'overlap-add' methods and those performed by using a time-domain implementation of equation (7.29). Suppose a long data sequence of length $M$ real points is to be processed by convolution with a real impulse response of length $L$:

$$M \gg L.$$

Employing equation (7.29) we require:

$$ML$$

real multiplications.

```
      SUBROUTINE FILTR(X,LEN)
C     DIMENSIONS OF ARRAYS MUST BE X1(2,M),X2(2,M),H(2,M),
C     AX1(2*M),AX2(2*M) AND AH(2*M) WHERE M = 2**(R - 1) AND
C     R IS SUITABLY CHOSEN, IN THIS CASE M = 64 SO THAT R = 7
      DIMENSION X(1),X1(2,64),X2(2,64),H(2,64),AX1(128),AX2(128)
      COMMON AH(128)
      EQUIVALENCE (AX1(1),X1(1,1)),(AX2(1),X2(1,1)),(AH(1),H(1,1))
      INTEGER R
C     SPECIFY VALUE OF R SO THAT LENGTH OF ARRAY AH IS 2**R
      R = 7
C     SPECIFY L IN ACCORDANCE WITH TABLE 1
      L = 24
      N = 2**R
      N1 = N/2
      R = R - 1
      A = 2.0 + FLOAT(LEN)/FLOAT(N-L+1)
      LIM1 = IFIX(A)
      CALL RFT3(H,R)
      DO 5 J = 1,LIM1
      DO 2 I = 1,N
      LIM3 = J*(N-L+1) + 1 - I
      LIM4 = N - 1 + 1
      IF(LIM3.LE.0.OR.LIM3.GT.LEN) GO TO 1
      AX1(LIM4) = X(LIM3)
      GO TO 2
    1 AX1(LIM4) = 0.0
    2 CONTINUE
      LIM5 = (J-2)*(N-L+1)
      LIM2 = N - L + 1
      DO 3 I = 1,LIM2
      IF(J.EQ.1) GO TO 3
      LIM6 = I + LIM5
      LIM7 = 1 + L - 1
      IF(LIM6.GT.LEN) GO TO 3
      X(LIM6) = AX2(LIM7)
    3 CONTINUE
      IF(J.EQ.LIM1) RETURN
      CALL RFT3(X1,R)
      X2(1,1) = X1(1,1)*H(1,1)
      X2(2,1) = X1(2,1)*H(2,1)
      DO 4 I = 2,N1
      X2(1,I) = X1(1,I)*H(1,I) - X1(2,I)*H(2,I)
      X2(2,I) = X1(1,I)*H(2,I) + X1(2,I)*H(1,I)
    4 CONTINUE
      CALL RFT4(X2,R)
    5 CONTINUE
      RETURN
      END
```

**Figure 7.25** Programme to perform the 'select-save' convolution on real data stored in an array X of length 'LEN'

If we employ the transform method, reference to Figure (7.24) indicates that:

$$N - L + 1$$

data points are processed on each transformation and weighting. Thus:

$$\frac{M}{N - L + 1}$$

transformations and weightings, in all, will be needed to obtain the convolution of all $M$ data points. Since each transformation is $N$ points long and real, and since we require two transformations plus one real $N$-point weighting for each 'pass', a total of:

$$\frac{M}{N - L + 1} \cdot [2N \cdot \log_2 (N) + 1]$$

real multiplications will be required to establish the convolution of the data by using the transform method. Inspection of Table 7.1 indicates, for the smaller values of $N$, that:

$$L \approx \frac{N}{5}.$$

Consequently we may express the number of real multiplications required for the transform method as approximately:

$$\frac{5M}{4} \cdot [2 \cdot \log_2 (N) + 1]$$

and for the time domain convolution as:

$$\frac{MN}{5}$$

A comparison of these two expressions indicates that the transform method is faster if:

$$N = 128.$$

Furthermore, if $N$ is much larger than this, a substantial speed increase results. For example, if

$$N = 4096$$

corresponding to values of $L$ in the range:

$$311 \leqslant L \leqslant 575$$

the transform method is about thirty times faster than the direct method.

## 7.11 Power spectrum estimation

We have defined in Section 7.2.6 the power spectrum of a sequence $x(k)$ with a DFT $X(n)$ as the sequence $P(n)$ where:

$$P(n) = |X(n)|^2; \qquad n = 0, 1, \ldots, N - 1.$$

As a representation of a possibly infinite time series, this power spectrum suffers from the fault, which we have discussed briefly in Section 7.4, of being derived only from a finite segment of the data. We are accustomed to think of power spectra as being smoothed (averaged) over all time.

It is helpful to summarize our objectives in attempting to establish a power spectrum. We shall take as our specification that the measured spectrum shall:

(a) be lowpass, or baseband, in nature
(b) have some desired resolution
(c) possess some required degree of statistical stability.

(a) The first of these conditions is reasonable, since spectra which are not baseband (and that usually means 'bandpass') may be represented by baseband spectra by applying suitable frequency translations. Now, in segmenting data from a long record, in order to perform the DFT required to establish the $A(n)$ and hence the $P(n)$, we introduce spurious spectral components. The segmentation process is equivalent to multiplying the original time series,

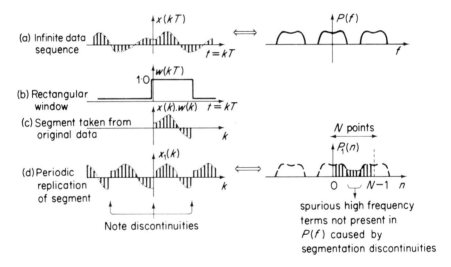

**Figure 7.26** Spurious high-frequency components caused by segmentation.

Figure 7.26(a), by a rectangular data 'window', Figure 7.26(b), passing $N$ data points, Figure 7.26(c). Consequently the 'true' spectrum of the time series is convolved with the Fourier Transform of the rectangular function. This results in a derived spectrum which can only approximate to the 'true' version, Figure 7.26(d).

The adverse effect of segmentation may be alleviated by the use of a data window which is other than rectangular. The basic requirement is that the window should cause the segmented time series to have no discontinuities at its endpoints. For example, one window function which has been used in power spectrum estimation, and which has the desirable feature of simplicity, both in its statement and implementation is:

$$w(k) = 0\cdot5[1 + \cos(2\pi k/(N - 1))]; \qquad k = 0, 1, \ldots, N - 1.$$

It can be shown that the interpolation process which corresponds to multiplication by this function gives modified spectral components $X_1(n)$ such that:

$$X_1(n) = -0\cdot25X(n - 1) + 0\cdot5X(n) - 0\cdot25X(n + 1).$$

Thus, instead of forming $P_{xx}(n)$ we should form $P_1(n)$ such that:

$$P_1(n) = |X_1(n)|^2; \qquad n = 0, 1, \ldots, N - 1$$

where:

$$x_1(k) = x(k) \cdot w(k); \quad k = 0, 1, \ldots, N - 1.$$

(b) The second condition, that the power spectrum should have particular resolution, may be taken to affect us in the following way. We have a sampled time function for which the sampling interval will have been established. We must, if we are not to run the risk of invalidating the sampling theorem, which we presume was used in setting up the sampled time function, examine sequences of consecutive points. A sequence, $N$ points in length, will have a duration of $NT$ seconds, and can be used to derive a spectrum specified with a frequency interval of $1/NT$ Hz. Since $T$ is fixed, we may therefore choose $N$ to suit some convenient criterion of resolution.

(c) The third requirement, that of statistical stability, refers to the fact that segments of data, each of length $N$, are taken from the same original sequence, $P_1(n)$. Statistical stability is, then, a criterion separate from resolution and segmentation. One solution to this problem is simply to average many spectra resulting from a succession of $N$-point segmentations. This process is illustrated in Figure (7.27). The long data sequence is broken into abutting $N$-point segments and the power spectrum of each segment is calculated. As we increase the number of contributions to the averaged spectrum its line components

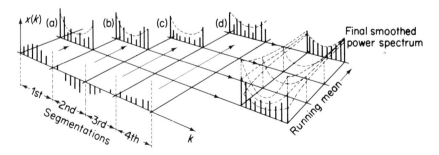

(a) (b) (c) and (d) are the power spectra resulting from the 1st 2nd 3rd and 4th segmentations at the time series, respectively.

**Figure 7.27.** Smoothing of power spectra.

```
      SUBROUTINE WKIN(R)
C     DIMENSIONS OF ARRAYS MUST BE AX(2,M/2),BX(2,M/4)
C     AND X(M) WHERE M = 2**R AND R IS SUITABLY CHOSEN
C     TO DETERMINE THE LENGTH OF THE DATA STRING TO
C     BE ANALYSED.
C     THE EQUIVALENCE STATEMENT MUST HAVE THE FORM:
C     EQUIVALENCE (X(1),AX(1,1)),(X(M/2 + 1),BX(1,1))
C     THE POWER SPECTRUM IS RETURNED IN THE FIRST HALF OF X
C     THE A.C.F. IS RETURNED IN THE SECOND HALF OF X
C     THE POWER SPECTRUM IS A REAL M/2 POINT FUNCTION
C     THE A.C.F. IS A REAL M/4 POINT FUNCTION
      DIMENSION AX(2,64),BX(2,32)
      COMMON X(128)
      EQUIVALENCE (X(1),AX(1,1)),(X(65),BX(1,1))
      INTEGER R
      R = R - 1
      N = 2**R
      N1 = N/2
      CALL RFT3(AX,R)
      DO 1 I = 1,N
      X(I) = AX(1,I)**2 + AX(2,1)**2
    1 CONTINUE
      DO 2 I = 1,N
      X(1 + N1) = X(I)
    2 CONTINUE
      R = R - 1
      CALL RFT4(BX,R)
      RETURN
      END
```

**Figure 7.28**  Programme to compute the power spectrum and autocorrelation function
of a real time series

exhibit a progressively smaller variance from the 'ideal' result, which would be obtained if we were able to allow the spectrum to average for 'all time'.

It should be noted that increasing $N$ will not, of itself, render the spectrum more stable, in the sense of decreasing the variance of the computed spectrum about the 'ideal'. This is because, although a greater time is devoted to establishing the spectrum by taking a longer record, more spectral points are calculated.

The subject of power spectral analysis has occupied a number of workers and the reader is directed, for a detailed account of the background to the subject, to Bingham[9], Welch[10] and Richards[11].

A simple subroutine, WKIN, which efficiently computes both the power spectrum and the autocorrelation function of a real time series is shown in Figure 7.28.

### References

1. Bracewell, R., *The Fourier Transform and its Applications*, McGraw-Hill, New York, 1965.
2. Singleton, R. C., 'A short bibliography of the fast Fourier transform,' *IEEE Trans Audio and Electroacoustics*, **AU-17** No. 2, 166 (1969).
3. Bergland, G. D., 'Fast Fourier transform hardware', *IEEE Trans Audio and Electro-acoustics*, **AU-17** No. 2, 166 (1969).
4. Cooley, J. W., and Tukey, J. W., 'An algorithm for the machine computation of complex Fourier series', *Math. of Comp.*, **19**, 297–301 (1965).

5. Brigham, E. O., and Morrow, R. E., 'The fast Fourier transform', *IEEE Spectrum*, **4**, 63–70 (1967).

6. Cochran, W. T., *et al.*, 'What is the fast Fourier transform?', *Proc. IEEE*, **55**, 1664–1674 (1967).

7. Singleton, R. C., 'A method of computing the fast Fourier transform', *IEEE Trans. Audio and Electroacoustics*, **AU-15** 91–98 (1967).

8. Helms, H. D., 'Fast Fourier transform method of computing difference equations and simulating filters', *IEEE Trans. Audio and Electroacoustics*, **AU-15** 85–90 (1967).

9. Bingham, C., Godfrey, M. D., and Tukey, J. W., 'Modern techniques of power spectral estimation', *IEEE Trans. Audio and Electroacoustics*, **AU-15** 56–65 (1967).

10. Welch, P. D., 'The use of the fast Fourier transform for the estimation of power spectra', *IEEE Trans. Audio and Electroacoustics*, **AU-15** 70–73 (1967).

11. Richards, P. I., 'Computing reliable power spectra', *IEEE Spectrum*, **14** 83–90 (1967).

## Chapter 8

# Frequency Sampling Filters

*R. E. Bogner*

## 8.1 Introduction

This technique provides filters which are very convenient to specify in the frequency domain, convenient to programme for simulation purposes and normally have truly linear phase characteristics. More storage is needed than for most recursive filters, but this may be shared among several filters in many applications.

There is a close correspondence between these filters and filters using the discrete Fourier transform. The impulse response is likewise of finite duration, like non-recursive filters.

The development highlights the convenience of complex number arithmetic available in digital systems.

## 8.2 Principle

The principle is relevant for continuous as well as digital filters; it is more practical for digital filters.

Consider sampling in the time domain. A continuous signal, bandlimited to $\pm W$ Hz may be reconstructed exactly from time samples taken at intervals of $1/2W = T$ seconds. The ideal interpolating filter has a frequency response constant from $-W$ to $W$ Hz and zero elsewhere. The corresponding impulse response is

$$h(t) = 2W \frac{\sin \pi t/T}{\pi t/T}.$$

Each of the samples becomes the amplitude of such a response. Frequency sampling filters (fsf's) use corresponding relations in the opposite domains, i.e. an impulse response time limited to $\pm \tau/2$ seconds, is represented by frequency samples at intervals of $1/\tau$ Hz. Figure 8.1 shows appropriate elemental frequency responses centred at 0, $1/\tau$ and $2/\tau$ seconds and the corresponding time responses:

$$e^{j(k2\pi t)/\tau}, \quad \left. -\frac{\tau}{2} < t < \frac{\tau}{2} \right\} \leftrightarrow \tau \frac{\sin \pi(f - k/\tau)\tau}{\pi(f - k/\tau)\tau} \qquad (8.1)$$

$$0 \qquad \text{elsewhere}$$

where the integer $k$ has the values 0, 1 or 2 in the illustrations.

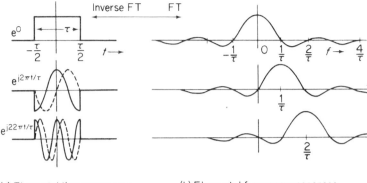

(a) Elemental time responses      (b) Elemental frequency responses

**Figure 8.1** (a) Elemental time responses; (b) elemental frequency responses

It is noted that each of these elemental frequency responses is zero at all sampling frequencies other than that at which it is one. This property ensures that a frequency response composed of the sum of such responses is influenced at each sampling frequency by only one of the specifying numbers—the value

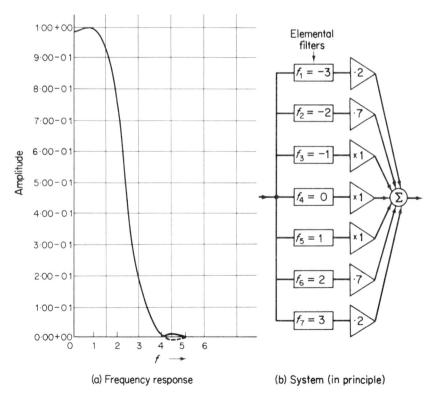

(a) Frequency response      (b) System (in principle)

**Figure 8.2** (a) Frequency response; (b) system (in principle)

of the relevant frequency sample. The resultant overall frequency response of the system is[1]

$$H(f) = \sum_k A_k \tau \frac{\sin \pi(f - k/\tau)\tau}{(f - k/\tau)\tau} \tag{8.2}$$

and the corresponding impulse response is

$$h(t) = \sum_k A_k e^{jk2\pi t/\tau} \tag{8.3}$$

for $-\tau/2 \leqslant t \leqslant \tau/2$.

A typical resultant response is shown in Figure 8.2(a), corresponding to the system of Figure 8.2(b). Each box represents a system with behaviour described by equation (8.1). In this example, the values of $A$ were:

$$k = -3 \quad -2 \quad -1 \quad 0 \quad 1 \quad 2 \quad 3$$

$$A = 0{\cdot}2 \quad 0{\cdot}7 \quad 1 \quad 1 \quad 1 \quad 0{\cdot}7 \quad 0{\cdot}2.$$

[These happen to be values which set the impulse response (8.3) to zero at $t = \pm\tau/2$.]

## 8.3  Realization of elemental responses

It is easiest to visualize the time response of a system in which the response is real, and mosts fsf's have used real elementary responses, $h_k(t)$:

$$h_k(t) = \begin{cases} \cos\dfrac{k2\pi t}{\tau}, & -\dfrac{\tau}{2} \leqslant t \leqslant \dfrac{\tau}{2} \\ 0 & , \quad \text{elsewhere.} \end{cases} \tag{8.4}$$

Figure 8.3 shows a 'comb filter' (see later for the name) whose impulse response is a unit impulse at $t = 0$, followed by a negative unit impulse at $t = \tau$. This filter is followed by a resonator whose impulse response is $\cos k2\pi t/T$, $t \geqslant 0$. As shown, the second impulse of the comb filter excites the cosine resonator in antiphase to the earlier excitation, yielding the time limited cosine impulse response for the combination.

The cosine responses, rather than being exponential, are simply related to the latter:

$$2\cos\frac{k2\pi t}{\tau} = e^{jk2\pi t/\tau} + e^{-jk2\pi t/\tau}.$$

Thus, if a cosine system is used, negative-frequency sample values are linked to positive-frequency values as is always the case for time responses which are real. The starting of the response at $t = 0$ rather than at $t = -\tau/2$ is necessary for realizability, and only corresponds to a pure delay in the whole system. All elemental responses must have the same delay $(\tau/2)$.

142

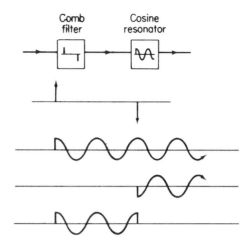

Production of elemental response

**Figure 8.3**  Production of elemental response

Since all the elemental responses are symmetrical about their centre points, the overall time response is symmetrical and the phase characteristic must be truly linear.

Figure 8.4 shows some suitable practical components. The cosine resonator is a second-order recursive filter with poles near the unit circle (z-plane). The complex number resonator[2] has an impulse response.

$$1, e^{(j\omega_k - \alpha)T}, e^{2(j\omega_k - \alpha)T}, e^{3(j\omega_k - \alpha)T}, \ldots,$$

(writing $\omega_k$ for $k2\pi/\tau$)

which may be seen by considering an applied impulse (unit sample), followed by zeros. Each successive circulation results in the multiplication of the output by $e^{(j\omega_k - \alpha)T}$. The damping, $\alpha$, is usually small and positive; it merely ensures stability in case of errors in the constants. The resultant transfer function of the $k$th resonator is

$$H_k(z) = \frac{1}{1 - z^{-1} e^{(j\omega_k - \alpha)T}}. \tag{8.5}$$

The complex number resonator is delightfully simple to programme—one line in Fortran:

$$Y = Y*B + X$$

where $X$ is the input, $Y$ the output and $B$ the value of $e^{(j\omega_k - \alpha)T}$. When only cosine responses are required, the real part of the output is taken.

The comb filter uses a considerable amount of storage; this may be unimportant in computer applications, and becomes less significant if the comb filter is shared among many resonators or filters in the complete filter.

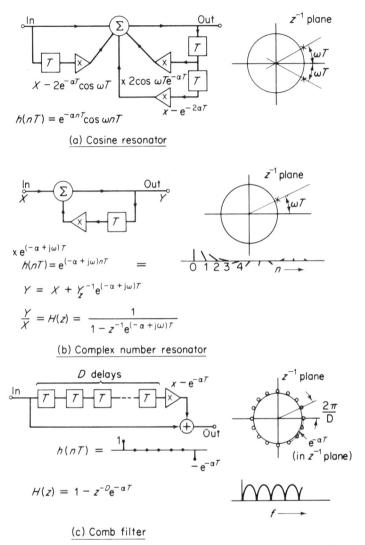

$$h(nT) = e^{-\alpha nT}\cos \omega nT$$

(a) Cosine resonator

$$x\,e^{(-\alpha + j\omega)T}$$
$$h(nT) = e^{(-\alpha + j\omega)nT}$$

$$Y = X + Y\,z^{-1}e^{(-\alpha + j\omega)T}$$

$$\frac{Y}{X} = H(z) = \frac{1}{1 - z^{-1}e^{(-\alpha + j\omega)T}}$$

(b) Complex number resonator

$$h(nT) = $$

$$H(z) = 1 - z^{-D}e^{-\alpha T}$$

(c) Comb filter

**Figure 8.4** (a) Cosine resonator; (b) complex number resonator; (c) comb filter

## 8.4 The complete filter

Figure 8.5 shows how one comb filter may be shared by many resonators to make one or several filters.

## 8.5 Pole–zero interpretation

The comb filter has zeros only, as is the case for all systems with finite duration responses. They occur where the pulse transfer function,

$$H(z) = 1 - z^{-D} = 0 \qquad (8.6)$$

where $DT = \tau$, the comb filter delay. Thus the zeros are uniformly spaced in frequency (hence the name 'comb' filter) at $z = \sqrt[D]{1}$ (Figure 8.4), i.e. at

$$z = e^{mj2\pi/D}, \qquad m = 0, 1, \ldots, D - 1. \tag{8.7}$$

For negligible damping, the poles of the (complex number) resonators occur at

$$z = e^{jk2\pi T/\tau} = e^{jk2\pi T/DT} = e^{jk2\pi/D} \tag{8.8}$$

(Figure 8.4) and thus cancel some of the zeros. It is just like nailing down the response, at regular intervals, and then pulling out some of the nails. The resultant zeros may be shown to occur in pairs on the same radius, cancelling each other's contribution; but the details are seldom important.

### 8.6 Relation to discrete Fourier transform[2] (DFT)

Consider $\alpha = 0$ and complex resonators. The response of the $k$th resonator at time $nT$, $n = 0, 1, 2, \ldots$, to a unit pulse at time $mT$ is, $e^{j\omega_k(n-m)T}$. Hence the $k$th response at time $nT$ to a signal $s(mT)$, $m = \cdots, -1, 0, 1, 2, \ldots, n$ is [Figure 8.5(a)]

$$x_k(nT) + jy_k(nT) = \sum_{m=-\infty}^{n} s(mT) e^{j\omega_k(n-m)T}$$

$$= e^{j\omega_k nT} \sum_{m=-\infty}^{n} s(mT) e^{-j\omega_k mT} \tag{8.9}$$

where $\omega_k = 2\pi k/DT = 2\pi k/\tau$.

When the comb filter precedes the resonator, the effect of its negative impulse, occurring $DT$ seconds after the positive impulse, is to add the second term of equation (8.10):

$$x_k(nT) + jy_k(nT) = e^{j\omega_k nT} \sum_{m=-\infty}^{n} s(mT) e^{-j\omega_k mT}$$

$$- e^{j\omega_k nT} \sum_{m=-\infty}^{n} s[(m-D)T] e^{j\omega_k mT}$$

$$= e^{j\omega_k nT} \left[ \sum_{m=-\infty}^{n} s(mT) e^{-j\omega_k mT} \right.$$

$$\left. - \sum_{m=-\infty}^{n-D} s(mT) e^{-j\omega_k mT} e^{-j\omega_k DT} \right]. \tag{8.10}$$

But $DT$ is an integral multiple of the period $2\pi/\omega_k$ and thus $e^{-j\omega_k DT} = 1$. Hence

$$x_k(nT) + jy_k(nT) = e^{j\omega_k nT} \sum_{m=n-D+1}^{n} s(mT) e^{-j\omega_k mT}. \tag{8.11}$$

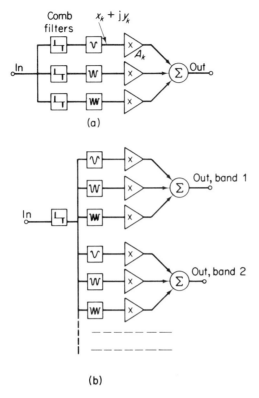

Figure 8.5

This expression may be recognized as an oscillation $e^{j\omega_k nT}$ whose coefficient is the value at frequency $\omega_k$ of the DFT of $s(mT)$, computed over the last $D$ samples. The output of the frequency sampling filter, taking into account the weights $A_k$, is thus

$$x(nT) + jy(nT) = \sum_k A_k[x_k(nT) + jy_k(nT)]$$

$$= \sum_k e^{j\omega_k nT}\left[ A_k \sum_{m=n-D+1}^{n} s(mT)\, e^{-j\omega_k mT} \right]. \qquad (8.12)$$

This is just the Fourier synthesis (inverse DFT) of the frequency function

$$A_k \sum_{m=n-D+1}^{n} s(mT)\, e^{-j\omega_k T}, k = 1, 2, \ldots \qquad (8.13)$$

which may be regarded as the product of the running DFT of $s(mT)$ and a DFT whose values at frequencies $\omega_k$ are the $A_k$.

*Frequency sampling filtering is thus equivalent to filtering by Fourier transforming, multiplying by a filter frequency function and inverse transforming.*

The filter frequency function ($A_k$, $k = 1, 2, \ldots$) has so far been considered to be real. There is no reason why the $A_k$ should not be complex, permitting the filter to have an arbitrary phase characteristic. The complex values of the $A_k$ may be specified in cartesian or polar form, the latter being more convenient for amplitude-phase specification. In this case, the system is no longer phase linear because the time response is not symmetrical.

## 8.7 Approximation due to sampling

We considered the elemental frequency responses of a continuous system (infinite sampling rate). The transfer function of a sampled data elemental filter is obtained by multiplying transfer functions of a complex resonator equation (8.5) and a comb filter equation (8.6):

$$H_k(z) = \frac{1 - z^{-D}}{1 - z^{-1} e^{j2\pi k/D}}$$

$$= \frac{1 - e^{j2\pi k} z^{-D}}{1 - e^{j2\pi k/D} z^{-1}}$$

since $e^{j2\pi k} = 1$. For the frequency response, we put $z = e^{j\omega T} = e^{j2\pi f T}$:

$$H_k(f) = \frac{e^{-j\pi(f - f_k)DT} \left[ e^{j\pi(f - f_k)DT} - e^{-j\pi(f - f_k)DT} \right]}{e^{-j\pi(f - f_k)T} \left[ e^{j\pi(f - f_k)T} - e^{-j\pi(f - f_k)T} \right]}$$

where $f_k = kDT$

$$\therefore \quad |H(f)| = \left| \frac{\sin \pi(f - f_k)DT}{\sin \pi(f - f_k)T} \right| \tag{8.14}$$

and the system is truly phase linear, with finite delay $DT/2$. The response magnitude (8.14) replaces the ideal equation (8.1), and we note that the significant difference lies in the denominator; when $\pi(f - f_k)T$ is small, i.e. the frequency difference from $f_k$ is small compared with $1/2T$ or the highest permitted frequency in the signal

$$\sin \pi(f - f_k)T \rightarrow \pi(f - f_k)T$$

and the functions behave similarly. This is the case for $T \rightarrow 0$, i.e. the continuous filter case.

The difference is entirely attributable to the repeated poles due to sampling; the denominator for the continuous case, $\pi(f - f_k)\tau$, contributes only one pole, at $f = f_k$.

## 8.8 Gibbs phenomena, shaping and sidelobes

Although the frequency response must always be as specified at the sampling frequencies, it may not be well behaved in between (Figure 8.6). The behaviour is related to the Gibbs phenomenon which describes the overshoot of a step

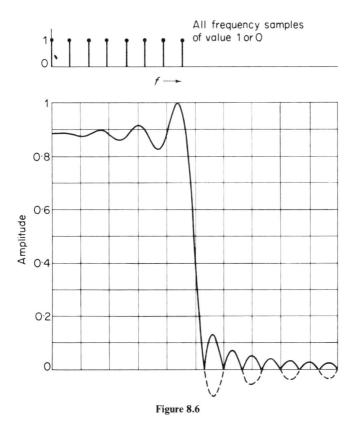

All frequency samples of value 1 or 0

**Figure 8.6**

function represented by a truncated (i.e. band limited) Fourier series. Appropriate weighting of the cutoff characteristics can result in smoother behaviour in pass- and stopbands [Figure 8.2(a)] but explicit formulae are not available. It is not hard to relate sidelobe suppression to the cancellation of tails of the 'sin $x/x$' responses, but criteria will vary from case to case. The earlier chapters on non-recursive filters and Fourier Transforms described the same problem in slightly different forms.

Recently[3], automatic optimization methods, in this case linear programming, have proven very powerful and convenient in proportioning the constants for sidelobe suppression. Chapter 11 introduces the general ideas of optimization.

### References

1. Rader, C. M. and Gold, B., 'Digital filter design techniques in the frequency domain', *Proc. IEEE*, **55** No. 2, 149 (1967).
2. Bogner, R. E., 'Frequency sampling filters—Hilbert transformers and resonators', *BSTJ* **48** No. 3, 501 (1969).
3. Rabiner, L. R., Gold, B., and McGonegal, C. A., 'The approximation problem for nonrecursive digital filters', Reported at IEEE Workshop on Digital Filtering, Harriman, N. J., Jan. 1970. (Probably to appear in *Trans. IEEE Audio and Electroacoustics*, June or Sept. 1970.)

*Chapter 9*

# Frequency-sampling Filters with Integer Multipliers

*P. A. Lynn*

## 9.1 Introduction

Previous chapters have already outlined the various ways in which a sampled-data signal may be filtered. In practice it is often difficult to choose the most appropriate technique for a particular application, and even when a decision has been made in favour of a time-domain operation it may be unclear whether to use a recursive or non-recursive filter. The amount of computation required to achieve a particular filtering action will generally be an important practical consideration; if a general-purpose computer is programmed as a digital filter, the computational economy of the filter design may well determine the feasibility of a real-time operation on the input data. But whether filtering is performed using a general-purpose computer or special-purpose hardware, it is generally the multiplication operations which are most expensive in terms of time and equipment. In other words, computational economy depends largely upon minimizing the number of multiplications required to calculate each new (filtered) output sample value. Furthermore, if the coefficients by which sample values are to be multiplied are small integers, multiplication will be far simpler than when those coefficients are floating-point numbers needing specification to an accuracy of perhaps five or six decimal figures.

In many cases the use of a recursive filter dramatically reduces the number of multiplications required, compared with a non-recursive filter having similar frequency-response characteristics. Unfortunately, the coefficients by which sample values must be multiplied in a recursive filtering operation must normally be specified with considerable accuracy. The reason for this is that a recursive design generally makes use of $z$-plane poles lying close to the unit circle, and a small error in the coefficient values of the time-domain recurrence equation may in effect cause these poles to move outside the unit circle, causing instability. (In this chapter, we are using the $z$-plane rather than the $z^{-1}$ plane, i.e. poles lie inside the unit circle for stability.) Apart from the problem of instability, the poles must normally be located with considerable accuracy if the required frequency-response characteristic is to be realized.

In this chapter, a family of recursive digital filters will be described in which all multiplier coefficients are small integers[1]. It will be shown that this practical advantage is only available if some rather severe restrictions on the locations of $z$-plane poles and zeros are accepted. These restrictions have the further advantage that all filters of the family display pure linear-phase characteristics,

150

imposing a pure transmission delay on all frequency components of an input signal. The filters to be described may in fact be considered as a special case of the Frequency-Sampling technique described in Chapter 8. Before describing them in detail, some general properties of linear-phase digital filters will be discussed.

## 9.2 Linear-phase digital filters

Any filter which has an impulse response (i.e. weighting function) symmetrical about $t = 0$ (the instant of application of the unit impulse) has a frequency response which is purely real. It may also be shown that a filter having an impulse response antisymmetrical about $t = 0$, in the sense that the portion of the response occurring before $t = 0$ is an inverted mirror image of the portion occurring after $t = 0$, has a purely imaginary frequency response; in other words it imposes a 90° or 270° phase shift at all frequencies. A simple example of a symmetrical weighting function is shown in Figure 9.1. Using methods similar to those explained in Chapter 3, page 38, the frequency characteristic is found to be:

$$H(j\omega) = 1 + e^{-j\omega T} + e^{j\omega T} = 1 + 2\cos \omega T$$

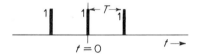

**Figure 9.1**  A simple symmetrical weighting function

which is purely real and whose magnitude is shown in Figure 9.2. It is thus clear that any weighting function which is symmetrical about $t = 0$ and is of finite duration yields a real frequency response equal to the sum of a number of

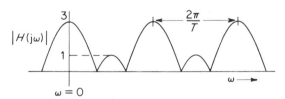

**Figure 9.2**  The frequency response magnitude function corresponding to the weighting function of Figure 9.1

cosines together with a constant term. Therefore the weighting function shown in Figure 9.3 has a frequency response of the form:

$$H(j\omega) = a_0 + 2a_1 \cos \omega T + 2a_2 \cos 2\omega T + \cdots 2a_n \cos n\omega T. \qquad (9.1)$$

A digital filter whose impulse response begins before $t = 0$ is not of course physically realizable, since it implies a response which precedes the input.

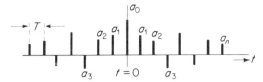

**Figure 9.3** A more complex symmetrical weighting
function

However, such a filter may be easily converted into a realizable one by shifting
the impulse response along the time axis so that it begins at or after $t = 0$.
This has no effect on the magnitude of the frequency-response, but converts
the zero-phase response of the original filter into a linear-phase one.

### 9.3 Recursive realization of linear-phase low-pass filters

The symmetrical weighting functions of the type described above may be
realized with computational economy when cast in recursive form. Further
advantages appear when the coefficients involved in the recursive form are
integers. This topic is conveniently introduced by referring to the simple 'moving
average' weighting function illustrated in Figure 9.4, which is symmetrical
about $t = kT$. Using $z$-transform notation, we may write the filter transfer
function directly:

$$H(z) = \frac{Y(z)}{X(z)} = 1 + z^{-1} + z^{-2} + \cdots z^{-2k}$$

$$- \frac{(1 - z^{-2k-1})}{(1 - z^{-1})} . \qquad (9.2)$$

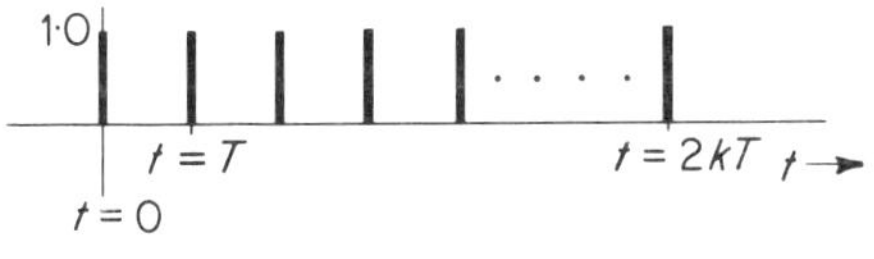

**Figure 9.4** A 'moving-average' weighting
function with $(2k + 1)$ terms

$H(z)$ has zeros where $(1 - z^{-2k-1}) = 0$, i.e. there are $(2k + 1)$ zeros evenly
distributed around the unit circle in the $z$-plane. There is also a single pole at
$z = 1$.

In the above equation $Y(z)$ and $X(z)$ are the output and input $z$-transforms
respectively, and we therefore obtain the relationship:

$$Y(z) = z^{-1} . Y(z) + (1 - z^{-2k-1}) . X(z),$$

which yields the time-domain recurrence formula:

$$y(n) = y(n - 1) + x(n) - x(n - 2k - 1). \qquad (9.3)$$

Equation (9.3) indicates that a recursive operation involving only 3 terms is equivalent to a non-recursive moving-average filter having any number of terms in its weighting function. For example if $k = 5$ the weighting function contains 11 terms and filtering is achieved using the recurrence formula:

$$y(n) = y(n-1) + x(n) - x(n-11).$$

The weighting function, pole-zero configuration and frequency response characteristic of this filter are illustrated in Figure 9.5.

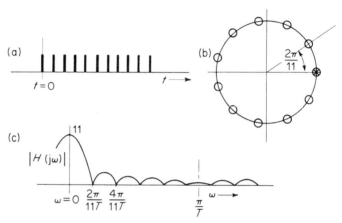

**Figure 9.5** 11-term 'moving-average' filter; (a) weighting function; (b) z-plane pole-zero configuration; (c) frequency-response magnitude function

Consider next the triangular weighting function shown in Figure 9.6. Using the result already derived (equation 9.2) and considering the triangular weighting function to be formed by the addition of a number of sets of unit-height samples, we may write directly:

$$H(z) = \frac{(1 - z^{-2k-1})}{(1 - z^{-1})} + \frac{z^{-1}(1 - z^{-2k+1})}{(1 - z^{-1})} + \cdots z^{-k}\frac{(1 - z^{-1})}{(1 - z^{-1})}$$

which reduces to

$$H(z) = \frac{(1 - z^{-k-1})^2}{(1 - z^{-1})^2}.$$

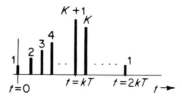

**Figure 9.6** A triangular weighting function with $(2k + 1)$ terms

The corresponding recurrence formula is therefore:

$$y(n) = -y(n - 2) + 2y(n - 1) + x(n - 2k - 2) - 2x(n - k - 1) + x(n). \quad (9.4)$$

This result shows that a triangular weighting function of any number of terms may be realized by a 5-term recursive filter. For example, $k = 10$ specifies a 21-term triangular weighting function. The $z$-plane transfer function of the filter may be written as:

$$H(z) = \frac{(1 - z^{-11})^2}{(1 - z^{-1})^2}.$$

Thus there are 11 second-order zeros spaced equally around the unit circle in the $z$-plane and a second-order pole at $z = 1$. The weighting function, pole-zero configuration and frequency response of this filter are illustrated in Figure 9.7. This filter may be realized by the recurrence formula:

$$y(n) = -y(n - 2) + 2y(n - 1) + x(n - 22) - 2x(n - 11) + x(n).$$

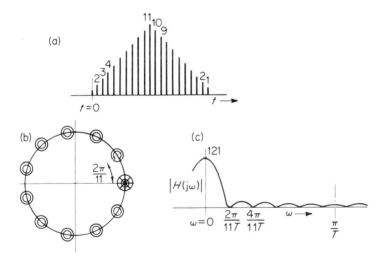

**Figure 9.7** 21-term 'triangular' filter: (a) weighting function; (b) $z$-plane pole-zero configuration; (c) frequency-response magnitude function

We notice that, with $k = 10$, the pole-zero pattern is similar to that of the simple moving-average filter previously discussed (with $k = 5$), except that double-zeros replace single zeros, and a double-pole replaces the single pole at $z = 1$. The transfer function of this 21-term triangular filter is thus the square of that of the 11-term moving-average filter. Recalling that multiplication in the frequency domain is equivalent to convolution in the time domain, we note that, as would be expected, the self-convolution of the moving-average weighting function does indeed yield the triangular one.

Both the filters so far discussed make use of $z$-plane zeros equally distributed around the unit circle, which are cancelled in one position by a coincident pole (or poles), giving rise to a passband. In general it may be shown that the placing of zeros at equal angular intervals on the unit circle, with cancellation by coincident poles in one or more positions, gives rise to recursive filters with the advantages of integer multipliers and linear-phase characteristics. It is always possible to raise a given transfer function $H(z)$ to an integer power; this has the effect of sharpening the cutoff and reducing the filter sidelobe levels.

## 9.4 Highpass and bandpass filters

The technique of placing cancelling poles on the unit circle at $z = 1$ may be modified to realize highpass and bandpass filters. Highpass filters are obtained by placing the poles at $z = -1$. As an example consider the transfer function

$$H(z) = \frac{(1 + z^{-7})^2}{(1 + z^{-1})^2}.$$

This may be realized by a 5-term recursive operation. The pole-zero configuration and weighting function are illustrated in Figure 9.8. When a highpass filter having a transmission zero at $\omega = 0$ is required, the numerator of $H(z)$ must be a term of the form $(1 + z^{-k})^n$, where $k$ is an even integer.

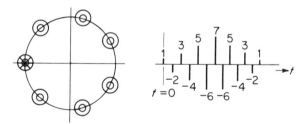

**Figure 9.8**  Pole-zero configuration and weighting function of a 5-term highpass recursive filter

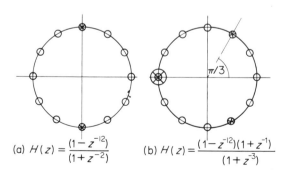

(a) $H(z) = \dfrac{(1 - z^{-12})}{(1 + z^{-2})}$  (b) $H(z) = \dfrac{(1 - z^{-12})(1 + z^{-1})}{(1 + z^{-3})}$

**Figure 9.9**  Pole-zero configurations of two bandpass filters: (a) centred on $\omega = \pi/2T$ (b) centred on $\omega = \pi/3T$

Bandpass filters may be obtained by placing cancelling poles at $z = \pm j$, giving a passband centred on the frequency $\omega = \pi/2T$, where $T$ is the sampling period. Passbands centred on $\omega = \pi/3T$ or $\omega = 2\pi/3T$ may also be obtained by placing three sets of cancelling poles at equal angular intervals around the unit circle, one of which is then made ineffective by the addition of further zeros. Figure 9.9 shows two typical pole-zero configurations which give bandpass characteristics.

### 9.5 Filter sidelobe levels

By increasing the order of poles and zeros we can reduce the sidelobe level compared with that of the main transmission lobe. However, the recurrence formula involves more terms, so that improved performance is paid for by increased computation. The sidelobe levels of a particular pole-zero configuration may be calculated by considering the lengths of vectors drawn from a point on the unit circle, (representing the frequency of interest) to the various poles and zeros. The calculation is considerably simplified if it is noted that the value of the response midway between two adjacent zeros is usually a reasonably good approximation to the peak value of the sidelobe in this interval. Using elementary geometry and trigonometry, it may be shown that the first sidelobes (which are the most significant) bear a ratio to the main lobe.

*9.5.1   Low- and highpass filters*

$$\frac{\text{Main lobe}}{\text{First sidelobe}} = \left[ k \sin \frac{3\pi}{2k} \right]^n$$

where $k$ — number of zeros distributed around the unit circle and $n = $ order of each zero. Note that $k^n$ is also the gain of the filter at the centre of the passband.

*9.5.2   Bandpass filters with centre frequency $\omega_c = \pi/2T$*

$$\frac{\text{Main lobe}}{\text{First sidelobe}} = \left[ \frac{k}{2} \sin \frac{3\pi}{k} \right]^n.$$

Note that $(k/2)^n$ equals the gain of the filter at the centre of the passband.

*9.5.3   Bandpass filters with centre frequency $\omega_c = \pi/3T$ or $2\pi/3T$*

$$\frac{\text{Main lobe}}{\text{First sidelobe}} = \frac{\left[ k \sin \left( \frac{3\pi}{2k} \right) \sin \left( \frac{\pi}{3} - \frac{3\pi}{2k} \right) \right]^n}{\sin^n \frac{\pi}{3}}$$

$$= \left[ 1 \cdot 15 \, k \sin \left( \frac{3\pi}{2k} \right) \sin \left( \frac{\pi}{3} - \frac{3\pi}{2k} \right) \right]^n.$$

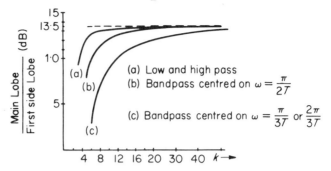

**Figure 9.10** The variation of sidelobe levels for 3 types of filter, as a function of the number of zeros ($k$) distributed around the unit circle in the $z$-plane

Figure 9.10 shows the locus of the above 3 functions as $k$ varies, for $n = 1$. If $n = 2$ (so that second-order poles and zeros are placed on the unit circle) the transfer function is the square of that for $n = 1$, and the decibel values given in Figure 9.10 must be doubled. If $n = 3$, they must be trebled, and so on.

### Reference

1. Lynn, P. A., 'Economic linear-phase recursive digital filters', *Electron. Lett.*, **6**, 143–145 (1970).

### Examples

*Frequency-sampling filters with integer multipliers*

1. Sketch the $z$-plane pole-zero configuration of a simple recursive 'moving average' lowpass filter having its first transmission zero at a frequency $\omega = \pi/6T$. Write down its transfer function and its time-domain recurrence formula. By considering the application of a single unit-height sample to the filter input, evaluate the filter's weighting function (i.e. impulse response).
2. A highpass filter has the transfer function:

$$H(z) = \frac{(1 - z^{-12})^2}{(1 + z^{-1})^2}.$$

Find its weighting function and hence use the convolution procedure to estimate its gain to the frequency $f = 1/2T$ Hertz.
3. Design a recursive bandpass filter with the following characteristics:
   (a) A transmission peak at a frequency $\omega = 2\pi/3T$.
   (b) A main-lobe width of $\pi/3T$ radians/second.
   (c) A first side lobe which is at least 17 decibels down on the main lobe.
   Write down its time-domain recurrence formula and estimate the saving in multiplication operations due to recursive design.

# Chapter 10

# Quantization Effects in Digital Filters

*V. B. Lawrence*

## 10.1 Introduction

The implementation of digital filters involves the use of finite numbers of bits to express their state and coefficient values. This fact produces several errors which are summarized below:

(1) quantization of input signal into a set of discrete levels;
(2) quantization of the filter coefficients into a finite number of bits;
(3) quantization of the results of arithmetic operations;
(4) low-level limit cycle oscillations (dead-band effects);
(5) overflow oscillations.

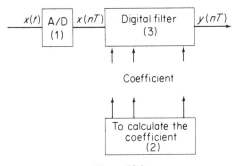

**Figure 10.1**

The occurrence of the first three errors are shown in Figure 10.1 and treated in detail in later parts of this chapter. The effects of these errors can be reduced arbitrarily by choosing the word lengths (numbers of bits) sufficiently large, but this increases the cost of the digital filter.

The degree of errors that are introduced into the operation of digital filters depends on the type of arithmetic operations used and the structure of the filter[1-4].

## 10.2 Realization

Four of the early types of realization, (1) direct, (2) direct canonic, (3) decomposed parallel and (4) cascade-series, are shown in Figures (10.17) to (10.20) in Appendix 10A.

## 10.3 Arithmetic considerations

The numerical operations in a digital filter are carried out using either fixed point or floating point arithmetic. The degree of error that is introduced in a filter operation using fixed point arithmetic is different from that of floating point. Fixed point arithmetic filters are easier to analyse but have the disadvantage of fixed binary points and more rigid limitations on their dynamic range[4].

### 10.3.1 Fixed-point arithmetic

In fixed point arithmetic, a state or coefficient variable, $U$ (less than unity), is represented in binary form as:

$$U = U_0 + \sum_{i=1}^{\infty} U_i 2^{-i} \qquad \text{where } U_i = 0 \text{ or } 1 \qquad (10.1)$$

and during implementation it is quantized to:

$$[U_q] = U_0 + \sum_{i=1}^{t-1} U_i 2^{-i} + \left[ U_t + \begin{Bmatrix} 1 \\ 0 \end{Bmatrix} \right] \qquad (10.2)$$

where with rounding a 1 or 0 is added to the $t$th bit depending on whether the $(t + 1)$th bit is a 1 or 0. With truncation, the bits beyond the most significant $t$-bits are simply dropped.

If the error that occurs during the quantization is denoted by $\varepsilon$,

$$\varepsilon = U - [U_q]$$

then for rounding $\varepsilon$ will lie between $-2^{-t}$ to $2^{-t}$, i.e.

$$-2^{-t} < U - [U_q] \leqslant 2^{-t}$$

and for truncation, the error will lie in the range $-2^{-t+1}$ to 0, i.e.

$$-2^{-t+1} < U - [U_q] \leqslant 0.$$

When two binary numbers $U$, $V$ or finite word lengths are multiplied, the resulting word will normally have a length greater than that of each of the original words. If a fixed number of bits per word is to be used throughout the implementation, the excess bits (at the least significant end) are usually truncated. This process results in an error which is similar to that obtained when a number is quantized by a quantizer with uniform step size. The addition of two binary numbers of finite word length does not result in error except when there is an overflow.

To study the accuracy problem of digital filters, caused by the quantization errors, it is necessary to know something about the properties of these errors. Most workers in this field[1-8] have found that the following three assumptions help produce theoretical results which are usually in close agreement with experimental results:

(1) the quantizing errors, $[U - [U_q]]$, can be regarded as random variables
(2) they are independent of each other and also of $U$ and $V$
(3) they have a uniform probability distribution in the range $|-2^{-t}$ to $2^{-t}|$
for rounding.

### 10.3.2 Floating-point

The errors in fixed point implementation have a constant range whilst in floating point implementation[3,6,9] they are proportional to the signal magnitudes. A floating point number is usually represented as shown in Figure 10.2.

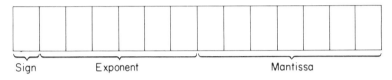

Sign          Exponent          Mantissa

**Figure 10.2**

In numerical form, the number $U$ is expressed as:

$$[\text{sgn}] . [2^{\alpha}][\beta] \tag{10.3}$$

where $\alpha$, the exponent is an integer which is greater than or equal to $\log_2 |U|$, and the mantissa $\beta$ is the result of the signal magnitude divided by the exponent:

$$\beta = U/[\log_2 U]. \tag{10.4}$$

The mantissa is the part which is always quantized and lies between 1 and $\frac{1}{2}$. Overflow occurs in the exponent (which determines the dynamic range of the filter).

For floating point arithmetic, both the sum and product of two finite word-length numbers have to be quantized. Since this quantization introduces errors into the system, the sum and product of two numbers are represented by;

$$\text{Sum}: [U + V]_q = (U + V)(1 + \varepsilon) \tag{10.5}$$

$$\text{Product}: [U . V]_q = (U . V) . (1 + \gamma) \tag{10.6}$$

where $\varepsilon$ and $\gamma$ are regarded as random variables which arise from either rounding or truncation and are taken to lie in the range;

$$-2^{-t} < \varepsilon, \quad \gamma < 2^{-t} \quad \text{for rounding}$$

$$-2^{-t} < \varepsilon, \quad \gamma < 0 \quad \text{for truncation.}$$

*The accuracy problem for the digital filters is analysed only for the fixed point case in this chapter.* The next subsection gives a detailed treatment of the various errors that occur in the different parts of the filter.

## 10.4 Quantization Effects

### 10.4.1 Quantization of the input

A continuous signal, $x(t)$, to be processed by a digital filter must be firstly digitized in amplitude and time. The process of digitizing each sample of the signal $x(nT)$ in amplitude involves a quantization into one of several levels, $E_0$, which normally introduces an error into the system. The effect of this error is analysed by considering it as additive noise superimposed at the input of the filter. The input to the filter is therefore considered as composed of two different sources.

(1) Noiseless input $x(nT)$ and (2) Additive noise $e(nT)$:

$$x_N(nT) = x(nT) + e(nT) \tag{10.7}$$

$$\underset{\substack{\text{Noise} \\ \text{less} \\ \text{input}}}{} \quad \underset{\substack{\text{Additive} \\ \text{noise}}}{}$$

Reasonable assumptions which can be made about the properties of this noise, $e(nT)$, are:

(1) the error associated with each sample is uniformly distributed in the range

$$-\frac{E_0}{2} \text{ to } \frac{E_0}{2} \text{ for rounding and } -E_0 \text{ to } 0 \text{ for truncation}$$

(2) the error at each sampling instant is statistically independent of the error at any other sampling instant.

The assumptions indicate that the noise is white, with zero mean for rounding and a variance, $\sigma^2$, equal to $E_0^2/12^{10-14}$. With these assumptions the variance or average noise power of the output error sequence can be determined by using linear system noise theory.

Considering the linear system shown in Figure 10.3 with transfer function $H(z)$; the output noise sequence $e_0(nT)$ is obtained by convolving $h(mT)$ with $e_{in}(nT)$.

$$e_0(nT) = \sum_{m=0}^{\infty} h(mT)e_{in}[(n-m)T]. \tag{10.8}$$

The sequence $e_0(nT)$ expressed by equation (10.8) can be regarded as a weighted sum of individual random variables with an autocorrelation function given by:

$$R_{00}(rT) = \sum_{n=0}^{\infty} e_0(nT) \cdot e_0[(n-r)T]$$

$$R_{00}(rT) = \sum_{n=0}^{\infty} \sum_{m=0}^{\infty} h(mT) \cdot e_{in}[(n-m)T] \cdot h[(m-r)T] \cdot e_{in}[(n-m-r)T]$$

| $e_{in}(nT)$ | Linear filter | $e_0(nT)$ |
| --- | --- | --- |
| Input white noise | $H(z)$ | Output |

**Figure 10.3**

which reduces after summing first over the index $n$ to

$$R_{00}(rT) = R_{in}(rT) . \sum_{m=0}^{\infty} h(mT) . h[(m-r)T]. \tag{10.9}$$

The variance (average output noise power) is given when $r = 0$ in equation (10.9), i.e.

$$R_{00}(0) = \sigma_0^2 = R_{in}(0) \sum_{m=0}^{\infty} h^2(mT)$$

and with $R_{in}(0)$ denoted by $\sigma_{in}^2$ or $E_0^2/12$ (see Appendix 10B) the above equation simplifies to:

$$\sigma_0^2 = \frac{E_0^2}{12} \sum_{m=0}^{\infty} h^2(mT). \tag{10.10}$$

When the impulse response $h(nT)$ is infinite, it is simpler to evaluate the output variance in the $z$-domain by calculating the residues of a contour integral defined from the discrete Parseval's theorem as:

$$\sigma_0^2 = \frac{E_0^2}{12} \sum_{m=0}^{\infty} h^2(mT) = \frac{E_0^2}{12} \cdot \frac{1}{2\pi j} \oint H(z)H^*(z)\frac{dz}{z} \tag{10.11}$$

(refer to Appendix 10.C). Use of the above formula is shown below in Example 1.

**Example 1.** Consider the first-order system shown in Figure (10.4) whose difference equation is given by:

$$y(nT) = a_1 y(nT - T) + x(nT) + e(nT)$$

where $(|a| < 1)$. The impulse response of the digital filter is given by

$$h(nT) = a_1^n. \tag{10.12}$$

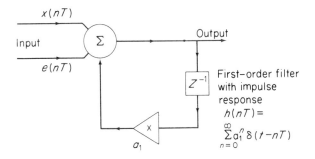

**Figure 10.4**  First-order filter with impulse response
$$h(nt) = \sum_{n=0}^{\infty} a_1^n \delta(t - nT)$$

---

* $H(z)$ is the transfer function of the ideal filter but could be modified to include round-off errors that occur after multiplication and also errors that arise from quantizing the coefficients.

Substituting equation (10.12) into (10.10) yields

$$\sigma_0^2 = \frac{E_0^2}{12} \sum_{n=0}^{\infty} a_1^{2n} \tag{10.13}$$

$$= \frac{E_0^2}{12(1 - a_1^2)}. \tag{10.14}$$

Using the $z$-domain analysis we have:

$$H(z) = \frac{1}{(1 - a_1 z^{-1})} \tag{10.15}$$

$$H^*(z) = \frac{z^{-1}}{(z^{-1} - a_1)} \tag{10.16}$$

and

$$H(z) \cdot H^*(z) = \frac{1}{(1 - a_1 z^{-1})} \cdot \frac{z^{-1}}{(z^{-1} - a_1)}. \tag{10.17}$$

Substituting (10.17) into (10.11) results in

$$\sigma_0^2 = \frac{E_0^2}{12(2\pi j)} \oint \frac{z^{-1}}{(z^{-1} - a_1)(1 - a_1 z^{-1})} \frac{dz^{-1}}{z^{-1}}. \tag{10.18}$$

Equation (10.18) is evaluated by calculating the residue due to the singularities within the contour of integration (unit circle). The singularities (poles) of $H(z) \cdot H^*(z)$ occur at $z^{-1} = a_1$ and at $z^{-1} = 1/a_1$ on the $z^{-1}$ plane. For $|a_1|$ less than unity there is only one pole within the contour of integration and this is at $z^{-1} = a_1$. The residue due to this pole is equal to $1/(1 - a_1^2)$. hence

$$\sigma_0^2 = \frac{E_0^2}{12(2\pi j)} \cdot 2\pi j \sum (\text{residue}) = \frac{E_0^2}{12(1 - a_1^2)} \tag{10.19}$$

and it is identical to equation (10.14).

If for example $a_1 = 0.9$ then

$$\sigma_0^2 = \frac{E_0^2}{12} \cdot 10^2.$$

Figure 10.5 indicates the variance (average output noise power) of three simple filters. Note in these examples the slight reduction in the output noise power (variance) when a zero is introduced. Hence the effect of a zero is to attenuate the noise.

### 10.4.2 Quantization of the filter into a finite number of bits

The ideal output sequence of a digital filter given by the difference equation;

$$h(nT) = \sum_{i=0}^{N} c_i x[(n - i)T] - \sum_{i=0}^{L} b_i h[(n - i)T] \tag{10.20}$$

| Type | Pole position | Approximate variance |
|---|---|---|

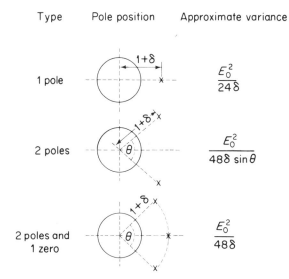

**Figure 10.5**

is altered if the coefficients $c_i$ and $b_i$ are quantized. The effect of this quantization is to replace the ideal coefficients, $c_i$ and $b_i$, by a finite number of bits. This gives rise to errors which are analogous to that obtained in the design of analogue filters, when the value of an inductance required might be 5·498 mH but when the coil is wound and the inductance carefully measured, its value would be 5·52 mH.

Quantization of the coefficients $c_i$ and $b_i$ changes the original pole and zero positions of the synthesized filter, on the $z^{-1}$-plane, thereby altering its amplitude and phase spectra. In certain circumstances, this error can cause stable filters to become unstable.

As a result of this quantization, the output sequence of a digital filter is expressed by;

$$[h_q(nT)] = \sum_{i=0}^{N} [c_i]_q x[(n-i)T] - \sum_{i=0}^{M} [b_i]_q \cdot h[(n-i)T] \qquad (10.21)$$

where

$$[c_i]_q = c_i + \gamma_i$$

and

$$[b_i]_q = b_i + \delta_i$$

both $\gamma_i$ and $\delta_i$ lie in the range $-2^{-t} < \gamma_i, \delta_i > 2^{-t}$ for fixed point arithmetic[6,9].

The transfer function representing such a 'quantized filter' is

$$[H_q(z)] = \frac{\displaystyle\sum_{i=0}^{N} [c_i]_q z^{-i}}{1 + \displaystyle\sum_{i=1}^{M} [b_i]_q z^{-i}}. \qquad (10.22)$$

A second-order system is used to illustrate the point that changes in coefficients alter the positions of the poles and zeros of the system.

**Example 2.** Consider the second-order difference equation

$$y(nT) = a_0 x(nT) + a_1 x(nT - T) + a_2 x(nT - 2T) + b_1 y(nT - T)$$
$$+ b_2 y(nT - 2T). \qquad (10.23)$$

Equation (10.23) has a system transfer function

$$H(z) = \frac{a_0 + a_1 z^{-1} + a_2 z^{-2}}{1 - b_1 z^{-1} - b_2 z^{-2}} \qquad (10.24)$$

with pole positions at

$$r_1 = \frac{b_1}{2} + \sqrt{\frac{b_1^2}{4} + b_2} \qquad (10.25)$$

$$r_2 = \frac{b_1}{2} - \sqrt{\frac{b_1^2}{4} + b_2} \qquad (10.26)$$

and zeros at

$$r_3 = -\frac{a_1}{2a_2} + \frac{1}{2a_2}\sqrt{a_1^2 - 4a_0 a_2} \qquad (10.27)$$

$$r_4 = -\frac{a_1}{2a_2} - \frac{1}{2a_2}\sqrt{a_1^2 - 4a_0 a_2}. \qquad (10.28)$$

Changes in the coefficients $a_0$, $a_1$, $a_2$, $b_1$ and $b_2$ could change the values of $r_1$, $r_2$, $r_3$ and $r_4$ and this would alter the amplitude and phase (frequency) characteristics of the filter. In certain cases, poles originally situated very near the unit circle on the $z^{-1}$-plane in the stable region might move into regions of instability. Hence it is necessary to search for realizations in which changes in the coefficients alter only slightly the positions of the poles. An example to show that different realizations produce different sensitivities is given below.

**Example 3.** Consider the second-order system whose poles are located on the $z^{-1}$-plane as illustrated in Figure 10.6. The poles are situated at $z^{-1} = 1/r_1$ and $z^{-1} = 1/r_2$. The transfer function of a system with these poles can be written in either of the following forms:

$$H_1(z) = \frac{1}{(1 - r_1 z^{-1})(1 - r_2 z^{-1})} \qquad (10.29)$$

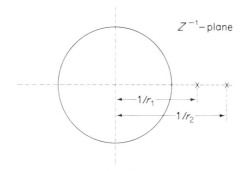

**Figure 10.6**

or

$$H_2(z) = \frac{1}{1 - (r_1 + r_2)z^{-1} + r_1 r_2 z^{-2}} \qquad (10.30)$$

The above transfer functions can be realized as follows:

(i) $H_1(z)$—Cascade (series) realization shown in Figure 10.7 with

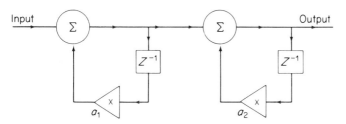

**Figure 10.7**  Cascade (series) realization

$$a_1 = r_1 \qquad (10.31)$$

$$a_2 = r_2. \qquad (10.32)$$

(2) $H_2(z)$—Direct form—illustration in Figure 10.8 with

$$b_1 = r_1 + r_2 \qquad (10.33)$$

$$b_2 = r_1 r_2. \qquad (10.34)$$

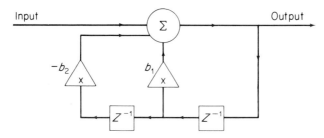

**Figure 10.8**  Direct-form realization

If equations (10.29) to (10.34) are partially differentiated we obtain

$$\frac{\partial r_1}{\partial a_1} = 1 \qquad (10.35)$$

$$\frac{\partial r_1}{\partial a_r} = 0 \qquad (10.36)$$

$$\frac{\partial r_2}{\partial a_1} = 0 \qquad (10.37)$$

$$\frac{\partial r_2}{\partial a_2} = 1 \qquad (10.38)$$

$$\frac{\partial r_1}{\partial b_1} = \frac{r_1}{r_1 - r_2} \qquad (10.39)$$

$$\frac{\partial r_1}{\partial b_2} = \frac{1}{r_2 - r_1} \qquad (10.40)$$

$$\frac{\partial r_2}{\partial b_1} = \frac{r_2}{r_2 - r_1} \qquad (10.41)$$

$$\frac{\partial r_2}{\partial b_2} = \frac{1}{r_1 - r_2} \qquad (10.42)$$

This shows that

$$\frac{\partial r_1}{\partial b_1} > \frac{\partial r_1}{\partial a_1} \qquad \text{if both } r_1 \text{ and } r_2 \text{ are of the same sign}$$

$$\frac{\partial r_1}{\partial b_2} > \frac{\partial r_1}{\partial a_2} \qquad \text{for all realizable values of } r_1 \text{ and } r_2$$

$$\frac{\partial r_2}{\partial b_1} > \frac{\partial r_2}{\partial a_2} \qquad \text{for all realizable values of } r_1 \text{ and } r_2$$

$$\frac{\partial r_2}{\partial b_2} > \frac{\partial r_2}{\partial a_2} \qquad \text{if both } r_1 \text{ and } r_2 \text{ are of the same sign.}$$

The above results indicate that for most practical filters, changes in $b_1$ and $b_2$ produce correspondingly larger variations in $r_1$ and $r_2$ than changes in $a_1$ and $a_2$. Hence in this example the direct-form realization appears to be more sensitive to quantization effects of the coefficients.

A generalized derivation of the relative movements in the positions of the poles due to coefficient variations can be deduced from a formula derived by J. Kaiser[15]. This formula which is simply quoted here in equation (10.43) is derived in Appendix 10D.

$$\Delta Z_r = \sum_{i=1}^{N} \left[ \frac{z_r^{i+1}}{\displaystyle\prod_{\substack{n=1 \\ n \neq r}}^{N} \left(1 - \frac{z_r}{z_n}\right)} \Delta b_i \right] \qquad (10.43)$$

where $\Delta b_i$ is the change in the value of the coefficient $b_i$ and $Z_1, Z_2, \ldots, Z_r$ are the positions of the poles of the ideal filter. From the above relationship, Kaiser proved that the sensitivity of changes in the pole positions with respect to the coefficients increases with increase in the order of filters realized in the direct form. Thus he concluded, it is not advisable to implement filters of order higher than the first or second, in the direct form. Therefore, the decomposed parallel and cascade forms are preferred.

### 10.4.3 Quantization of the results of arithmetic operations

The result of every multiplication of the coefficient and state variable has to be truncated or rounded. This operation on the products of multiplications introduces errors into the system at the various nodes[1] as each iteration of the difference equation is executed. The effect of this error which is similar to the quantization of the input is analyzed by considering white noise to be super-imposed at the various nodes within the filter. As this noise is superimposed at various nodes in the system, the average noise power at the output would depend on the structure of the filter.

There are basically four main ways in which a transfer function can be realized.

(1) Direct form.
(2) Direct canonic form (fewer delay elements).
(3) Parallel (direct and canonic forms).
(4) Cascade (direct or canonic forms).

The average output noise power due to the quantization of the results of arithmetic operation, is evaluated for each of the four structures of the digital filter in the next sub-sections.

### 10.4.3.1 Direct form.

A noise model for the direct form realization of an $N$ zero, $M$ pole filter is shown in Figure 10.9, with the noise sources inserted at various nodes within the filter. These noise sources all flow directly into the summing junction and so can be combined into a single noise source inserted

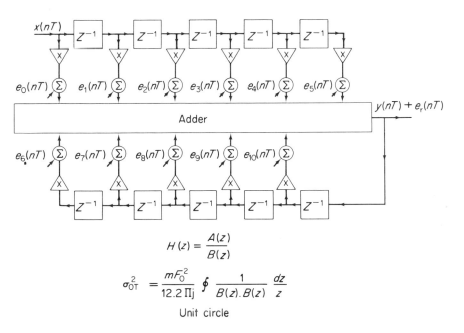

$$H(z) = \frac{A(z)}{B(z)}$$

$$\sigma_{OT}^2 = \frac{mF_0^2}{12.2\,\Pi j} \oint \frac{1}{B(z).B(z)} \frac{dz}{z}$$

Unit circle

**Figure 10.9**  Noise model for direct form

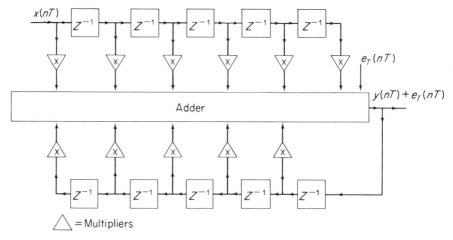

$x(nT)$

$e_T(nT)$

Adder

$y(nT) + e_T(nT)$

$\triangle$ = Multipliers

**Figure 10.10**

at the summing junction as illustrated in Figure 10.10. This simplified equivalent noise model clearly demonstrates that the noise passes only through the poles of the filter which are contributed by the $b_i$ coefficients. Thus the transfer function from the noise source to the output consists of only poles whereas the signal passes through the zeros (which are contributed by $c_i$ as well). The average noise power at the output of the filter is therefore calculated as follows:

Let the transfer function of the filter be represented by

$$H(z) = \frac{C(z)}{B(z)}$$

where $C(z)$ specifies the zeros and $B(z)$ the poles. Using the assumption that the noise source $e(nT)$ is zero mean and wide sense stationary[1-3,5-11] the power spectral density of the output noise will be given by

$$S_{00}(z) = S_{11}(z) \cdot \left[\frac{1}{B(z)}\right] \cdot \left[\frac{1}{B(z)}\right]^* \tag{10.44}$$

where $1/B(z)$ is the effective transfer function through which the noise passes and $S_{11}(z)$ is the power spectral density of the noise inserted at the summing junction.

The average output power of the noise is the mean square value of the noise sequence and this is equal to the mean of the autocorrelation function at the origin. The autocorrelation function of $e_0(nT)$ is given by:

$$R_{00}(r) = \sum_{n=0}^{\infty} e_0(nT) \cdot e_0[(n)T] \tag{10.45}$$

at the origin where $r = 0$ it is expressed as

$$R_{00}(0) = \sum_{n=0}^{\infty} e_0(nT) \cdot e_0(nT). \tag{10.46}$$

The autocorrelation function at the origin is the Fourier transform of the power spectral density function and therefore the mean of the autocorrelation function at the origin is equal to the mean of the power density function. i.e.

$$R_{00}(0) \xrightarrow{\text{FT}} S_{00}(z)$$

$$\overline{R_{00}(0)} = \frac{1}{2\pi j} \oint S_{00}(z) \frac{dz}{z}. \tag{10.47}$$

Thus from equations (10.45) and (10.47) the average output noise power will be given by;

$$\{E \, e_0(nT)\} = \sigma_{DT}^2 = (1 + N + M) \frac{E_0^2}{12(2\pi j)} \oint \frac{1}{B(z)} \frac{1}{B(z)} \frac{dz}{z} \tag{10.48}$$

where $(1 + N + M)E_0^2/12$ is the average power of the noise caused by rounding off the products of multiplication, $E_0$ is the quantization step and $M + N$ the number of poles and zeros.

10.4.3.2 *Canonic realization* The canonic realization uses half the number of delay elements needed for the direct realization. The noise and equivalent noise models of this realization shown in Figures 10.11 and 10.12 respectively, indicate that the combined noise source enters the system in a different way to that of the direct realization. The combined noise, $e_b(bT)$, which arises from the product of the $b_i$ coefficients and state variables* passes through the entire network (of poles and zeros) whereas the $e_c(nT)$ noise which is contributed by the multiplication of $e_i$ and the state variables is simply added noise at the output.

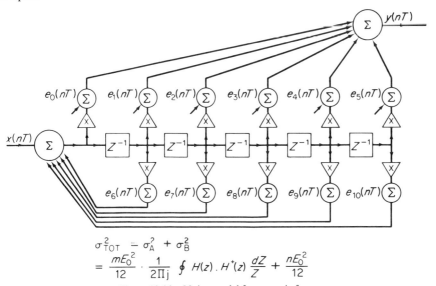

$$\sigma_{TOT}^2 = \sigma_A^2 + \sigma_B^2$$
$$= \frac{mE_0^2}{12} \cdot \frac{1}{2\Pi j} \oint H(z) . H^+(z) \frac{dZ}{Z} + \frac{nE_0^2}{12}$$

**Figure 10.11** Noise model for canonic form

* State variables are the time samples at the output of the nodes.

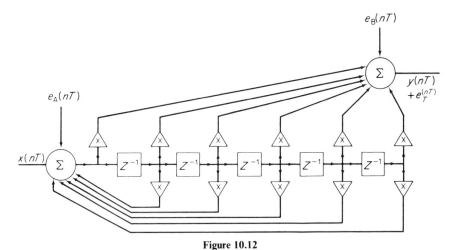

**Figure 10.12**

Thus the output variance (total average power) will be the sum of the individual output variances contributed by $e_b(nT)$ and $e_c(nT)$.

Output variance contributed by $e_b(nT)$ is:

$$E\{e_b(nT)\} = \sigma_{bT}^2 = \frac{ME_0^2}{12} \cdot \frac{1}{2\pi j} \oint H(z)H^*(z)\frac{dz}{z} \qquad (10.49)$$

where $H(z)$ is the effective transfer function through which $e_b(nT)$ passes and $M$ is the number of poles. The output variance due to $e_c(nT)$ is given by

$$E\{e_c(nT)\} = \sigma_c^2 = (1 + N)\frac{E_0^2}{12}. \qquad (10.50)$$

Therefore, the total output noise variance which is the sum of

$$E\{e_b(nT)\} \quad \text{and} \quad E\{e_c(nT)\}$$

which is expressed as

$$\sigma_T^2 = \frac{M \cdot E_0^2}{12 \cdot 2\pi j} \oint H(z) \cdot H^*(z)\frac{dz}{z} + (1 + N)\frac{E_0^2}{12}. \qquad (10.51)$$

Thus in the canonic realization, part of the noise and the whole of the signal pass through the zeros as well as the poles of the system. Generally the effect of zeros is to attenuate the noise, hence it appears that the canonic realization is better than the direct realization as far as fixed-point round-off arithmetic errors are concerned. This may not always be true (refer to example 5 in Appendix 10.E). E. K. Kan and J. K. Aggarwal[19] have shown that for floating point arithmetic, the degree of round-off noise generated within the filter is the same for both canonic and direct realizations.

*10.4.3.3  Parallel form*  For the parallel form of realization, shown in Figure 10.13, the transfer function is decomposed into a linear combination of first- or second-order transfer functions.
i.e.

$$H(z) = K + H_1(z) + H_2(z) \dots H_N(z)$$

$$= K + \frac{\alpha_1}{1 - \beta_1 z^{-1}} + \frac{\gamma \, \delta z^{-1}}{\pi + \xi z^{-1} + \eta z^{-1}}. \tag{10.52}$$

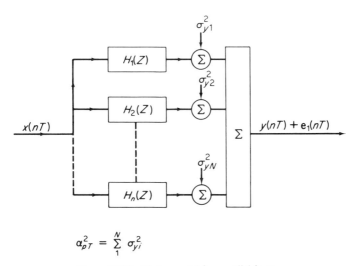

$$\alpha_{pT}^2 = \sum_1^N \sigma_{yi}^2$$

**Figure 10.13**  Noise model for parallel form

Each of these reduced first- or second-order transfer functions is either realized in the direct or canonic form. The noise variance $\sigma_i^2$, at the output of each section, $H_i(z)$, is computed using either equation (10.48) or (10.51) depending on the structure of the filter. The overall output noise variance is the sum of the individual average noise power at the output of each section, i.e.

$$\sigma_{pT}^2 = \sum_{i=1}^{L} \sigma_i^2 \tag{10.53}$$

where $L$ is the number of parallel paths.

*10.4.3.4  Cascade form*  The cascade form of realization shown in Figure 10.14 has a transfer function which is broken down into first- or second-order structures as:

$$H(z) = KH_1(z) \, . \, H_2(z) \dots H_r(z) \dots H_N(z)$$

$$= K \, . \, \frac{z^{-1} - \alpha_1}{1 - \beta_1 z^{-1}} \, . \, \frac{z^{-1} - \alpha_2}{1 - \beta_2 z^{-1}} \dots \frac{z^{-2} - \gamma z^{-1} + \delta}{\pi z^{-2} + \xi z^{-1} + 1}. \tag{10.54}$$

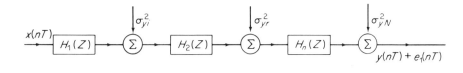

$$\sigma_{ST}^2 = \sum_{i=1}^{N} \sigma_{yi}^2 \cdot \frac{1}{2\Pi j} \oint \prod_{j=i+1}^{N} [H_j(z) \cdot H_j^*(z)] \cdot \frac{dz}{z}$$

**Figure 10.14** Noise level for cascade form

Likewise, the output noise power for each section is calculated using either equation (10.48) or (10.51) depending on the realization. For the cascade realization, the input noise at each section includes the noise at the output of the previous section as well as the noise generated within the current section. The output noise therefore from the first stage passes through the zeros and poles of the rest of the stages. In general the average output noise power at the $i$th stage passes through all the poles and zeros of the $(i + 1)$th and higher stages of the system.

In a mathematical form the total average noise power can be expressed as:

$$\sigma_{sT}^2 = \sum_{i=0}^{L} \sigma_i^2 \frac{1}{2\pi j} \oint \prod_{k=i+1}^{L} [H_k(z) \cdot H_k^*(z)] \frac{dz}{z} \tag{10.55}$$

where $L$ is the number of sections.

*10.4.3.5 Conclusion* In general the effects of round-off noise vary with the specific configuration chosen, even though all the configurations have the same overall transfer function. To choose a particular configuration which best minimizes the bad effects of noise, equations (10.48), (10.51), (10.53) and (10.55) would have to be evaluated. In most cases the cascade form is proved best[2].

## 10.5 Noise measurement

Figure 10.15 below is a simple experimental arrangement to measure the round-off noise generated within a digital filter. In this diagram there are two

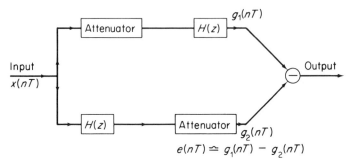

**Figure 10.15** Noise measurement

identical filters in parallel with transfer function, $H(z)$. An attenuator is placed before the filter in the first path and after the filter in the second path of the network. The same signal which is applied to both paths of the network is first attenuated along path one before passing through the filter, and along the second path the signal is attenuated after passing through the filter. In this arrangement the round-off noise generated within the first filter is not attenuated. Thus the difference between the signals at the output of the filter in path one and the output of the attenuator in path two gives a direct measure of the round-off noise generated within a filter with a transfer function $H(z)$.

## 10.6 Low-level limit cycles

The round-off and truncation errors that occur after multiplication at various nodes within the filter, and described in the previous sub-section, were assumed to be independent of each other and uncorrelated from sample to sample. These assumptions have been found by most workers[2-9,16-20] on this subject to produce theoretical results which are in close agreement with experimental results provided the input signal to the filter is not of low amplitude. For low-level input signals the round-off errors become correlated and, in the limit of zero input to the filter, periodic signals or limit cycles are observed at the output of recursive digital filters. Thus filters whose impulse response should decay to zero may not as a result of the quantization of the products of multiplication at various nodes within the filter. (Filters which are supposed to be asymptotically stable in the bounded input, bounded output sense become marginally stable.)

Low-level limit cycles errors do not produce instability, provided no overflow occurs[5]. The example 4 below is used to illustrate the occurrence of limit cycles in digital filters.

**Example 4.** Consider a first-order digital filter whose difference equation is given by

$$y(nT) = x(nT) - 0{\cdot}9[y(nT - T)] \qquad (10.56)$$

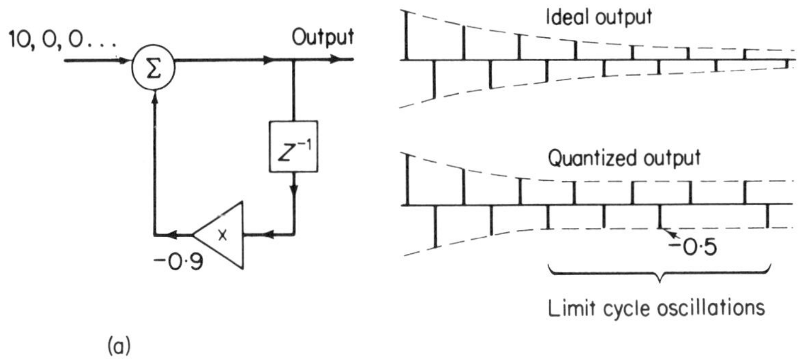

(a)

**Figure 10.16**(a)

$y(nT)$, $x(nT)$ and 0·9 are the output signal, input signal and the multiplier coefficient, respectively. A filter with the above difference equation can be realized as shown in Figure 10.16a.

For an input signal 10, 0, 0, ..., the ideal output signal will be as shown on the left-hand side of Table 10.1 below. If the result of each multiplication of $y(nT - T)$ and $-0.9$ is rounded-off to the nearest integer, the output samples will be as shown on the right-hand side of the same table. It can be seen from Table 10.1 that the filter starts oscillating after $5T$ seconds. The amplitude of oscillation is 5 and the frequency of oscillation is equal to $\frac{1}{2}$(sampling frequency) of the filter. Different orders of filter and multiplier coefficients produce different amplitudes and frequencies of oscillation.

Table 10.1

| Input samples—10, 0, 0, ... | | |
|---|---|---|
| Output samples | | |
| Time | Ideal | Quantized |
| $0T$ | +10 | +10 |
| $1T$ | −9 | −9 |
| $2T$ | +8.1 | +8 |
| $3T$ | −7·29 | −7 |
| $4T$ | +6·561 | +6 |
| $5T$ | −5·8949 | −5 |
| $6T$ | +4·80541 | +5 |
| $7T$ | −4·324869 | −5 |
| $8T$ | +3·8923821 | +5 |
| ⋮ | ⋮ | ⋮ |
| $\infty T$ | 0·000000 | ±5 |

## 10.7  Overflow oscillations

Overflow oscillations may occur after the execution of the difference equation at various nodes within the filter because of the finite dynamic range of the filter. These oscillations are very large and undesirable, therefore it is important to eliminate their occurrence. A simple method is to zero the outputs of all nodes where overflow has occurred. Other more sophisticated techniques have been developed by I. W. Sandberg[6] and P. M. Ebert et al.[21]. It is not all overflows that occur at the nodes that are harmful. L. Jackson et al.[22] have shown that some overflow occurrence during partial sums of two or more samples can be allowed.

**Example.** Let us consider an instant during the operation of an 8-bit digital filter, when we have to add the following numbers, 0·0, +5·0, +6·0, −2·0, −4·0.

Let us assume the fixed point binary representation of the state variables of the filter to be:

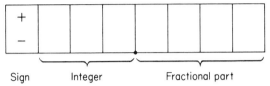

The first bit represents the sign, the next three the integer part and the last four the fractional part. The magnitude of the magnitude of the maximum number that can be represented is $7\frac{15}{16}$.

The partial sums using 2's complement arithmetic carried out during the addition of the above numbers is illustrated below in Table 10.2.

Table 10.2

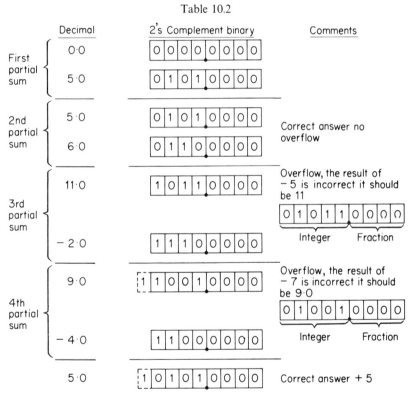

From Table 10.2 we see that there were two temporary overflows but the final result of $+5$ was correct. Thus overflows can be tolerated temporarily provided the magnitude of the total sum does not exceed the range of the filter (in this example $7\frac{15}{16}$). A diagrammatic representation of the execution of partial sums is shown in Figure 10.16b.

(1) First partial sum;
Starting from $A(0,0)$ and adding $5\cdot0$ one moves counterclockwise 5 steps to $B(5, 0)$.

176

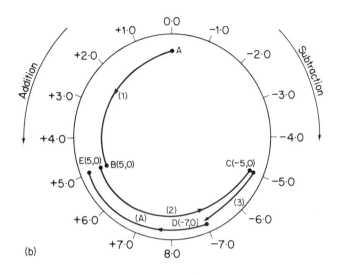

**Figure 10.16(b)**

(2) Second partial sum;
    Starting from point B(5, 0) one moves 6 steps to the point C(− 5, 0).
(3) Third partial sum:
    Starting from point C, one moves 2 steps in the clockwise direction to
    the point D(− 7, 0).
(4) Fourth partial sum:
    Starting from the point D(− 7, 0) one moves 4 steps in the clockwise
    direction to the point E (or B), which is + 5·0.

APPENDIX 10A

REALIZATION

Filters which have both poles and zeros are known as recursive or Infinite
Impulse Response (IIR) filters. Their realization is also in terms of adders,
multipliers and delay elements. They differ only from (FIR) filters in that they
have feedback elements (that is they have $(s − 1)$ feedback delay elements
and $(r − 1)$ feedforward elements). Four of the early types of realization are
shown in Figures 10.17 to 10.20. The first is the direct form illustrated in Figure
(10.17) with $(r + s)$ delay elements. In the direct form realization shown above,
the number of delay elements is greater than the order of the filter. It is possible
to realize the same transfer function using fewer delay elements. For example,
the number of delay elements needed to realize $H(z)$ could be $s$ if $s > r$ or $r$
if $s < r$. The larger of these two numbers is the order of the filter. Realizations
in which the number of delay elements equals the order of the transfer function
are referred to as *Canonic*. The transfer function

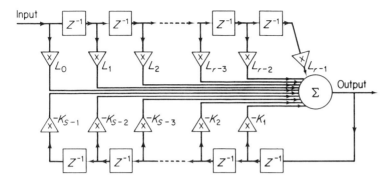

**Figure 10.17** Direct form realization

$$H(z) = \frac{\sum\limits_{i=0}^{r} L_i z^{-i}}{\sum\limits_{i=1}^{s} k_i z^{-i}} \tag{10.57}$$

is rewritten as

$$H(z) = W(z) \sum_{i=0}^{r} L_i z^{-i} \tag{10.58}$$

where

$$W(z) = \frac{1}{\sum\limits_{i=0}^{s} k_i z^{-i}}. \tag{10.59}$$

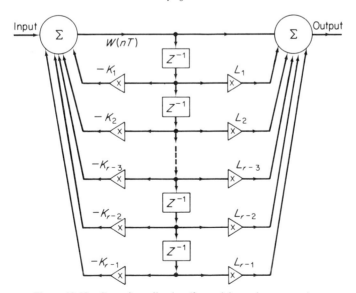

**Figure 10.18** Canonic realization (fewer delays—less storage)

The transfer function $H(z)$ can be further broken down into first- or second-order structures and realized as either a parallel or cascade combination of lower-order functions (see Figures 10.19 and 10.20).

Parallel Combination

$$H(z) = K + H_1(z) + H_2(z) \ldots H_N(z) \ldots \tag{10.60}$$

$$= K + \frac{\alpha_1}{(1 - \beta_1 z^{-1})} + \frac{\gamma_1 + \delta z^{-1}}{(\pi + \xi z^{-1} + \eta z^{-2})}. \tag{10.61}$$

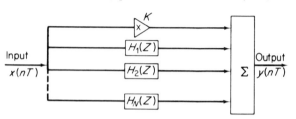

**Figure 10.19**  Parallel realization

Cascade (series)

$$H(z) = K H(z) . H_2(z) \ldots H_r(z) \ldots H_n(z) \tag{10.62}$$

$$\left[\frac{z^{-1} - \alpha_1}{z^{-1} - \beta}\right] \left[\frac{z^{-1} - \alpha_2}{z^{-1} - \beta_2}\right] \ldots \left[\frac{z^{-2} + \gamma z^{-1} + \delta}{z^{-2} + \xi z^{-1} + \pi}\right] \tag{10.63}$$

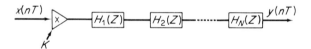

**Figure 10.20**

where each block $H_i(z)$ is realized in either the canonic or direct forms of Figures 10.17 or 10.18, respectively.

## APPENDIX 10B

### VARIANCE OF ERRORS DUE TO ROUNDING

The following assumptions were made about the round-off errors due to arithmetic operations

(1) Error, $\varepsilon$, is uniformly distributed in the range $-E_0/2$ to $+E_0/2$, where $E_0$ is the quantization step.
(2) The mean of $\varepsilon$ is zero.
(3) The errors, $\varepsilon$ are independent of each other and uncorrelated from sample to sample.

With these assumptions, the probability distribution, $p(\varepsilon)$, of the error is as shown in Figure 10.21 below.

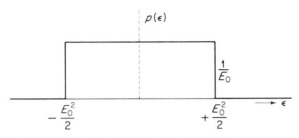

**Figure 10.21**  Probability distribution of round-off errors

$$p(\varepsilon) = \frac{1}{E_0}$$

by definition, the variance[23] of errors

$$\sigma_0^2 = \int_{-\infty}^{+\infty} (\varepsilon - \mu)^2 p(\varepsilon)\, d\varepsilon \qquad (10.64)$$

where $\mu$ is the mean and in this case, $\mu = 0$ and $p(\varepsilon) = 0$ for $\varepsilon > E_0/2$ and $\varepsilon < -E_0/2$. Under these conditions equation (10.64) becomes:

$$\sigma_0^2 = \int_{E_0/2}^{E_0/2} (\varepsilon - 0)^2 \frac{1}{E_0}\, d\varepsilon. \qquad (10.65)$$

Evaluation of the integral expressed in equation (10.65) gives

$$\sigma_0^2 = \frac{1}{E_0}\left[\frac{\varepsilon^3}{3}\right]_{-E_0/2}^{E_0/2}$$

$$= \frac{E_0^2}{12}.$$

## APPENDIX 10C

## DISCRETE PARSEVAL'S THEOREM

Consider a linear system shown in Figure 10.22, where $x(t)$ is the input time function, $y(t)$ the output time function and $h(t)$ the impulse response of the linear system (filter). The power spectrum of the output function $y(t)$, denoted by $S_{yy}(f)$ is given by[24]:

$$S_{yy}(f) = S_{xx}(f) \cdot |H(f)|^2. \qquad (10.66)$$

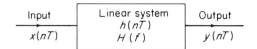

**Figure 10.22**

where $S_{xx}(f)$ is the power spectrum of the input function $x(t)$ and $H(f)$ is the frequency characteristics of the filter. The total average output power is the integral of the power spectrum and is defined as:

$$\sigma_0^2 = \int_{-\infty}^{\infty} S_{yy}(f)\,df. \tag{10.67}$$

From equations (10.66) and (10.67) the total average output power can be expressed as

$$\sigma_0^2 = \int_{-\infty}^{\infty} S_{xx}(f)H(f)H^*(f)\,df. \tag{10.68}$$

If $x(t)$ is assumed to be gaussian with a variance $\sigma_{in}^2$ then equation (10.68) simplifies to

$$\sigma_0^2 = \sigma_{in}^2 \int_{-\infty}^{\infty} H(f)H^*(f)\,df. \tag{10.69}$$

The above equation (10.69) can be transformed from the frequency domain into the $z$-domain by the use of the following transformations

$$z = e^{j2\pi f} \tag{10.70}$$

such a transformation maps the imaginary axis of the S-plane onto the unit circle in the Z-plane[25], and the integral, $\int_{-\infty}^{\infty}$, over the limits $-\infty$ to $+\infty$ changes into a contour integration, $\oint$, over the unit circle.

Differentiating equation (10.70) gives:

$$dz = j2\pi \cdot e^{j2\pi f} \cdot df \tag{10.71}$$

$$df = \frac{dz}{j2\pi z}. \tag{10.72}$$

Hence use of the transformation changes equation (10.69) into

$$\sigma_0^2 = \sigma_{in}^2 \cdot \frac{1}{2\pi j} \oint H(z) \cdot H^*(z)\frac{dz}{z}. \tag{10.73}$$

From equation (10.10)

$$\sigma_0^2 = \sigma_{in}^2 \sum_{m=0}^{\infty} h^2(mT). \tag{10.74}$$

Equating the right-hand sides of equations (10.73) and (10.74) results in Parseval's Theorem.

$$\sum_{m=0}^{\infty} h^2(mT) = \frac{1}{2\pi j} \oint H(z) \cdot H^*(z) \frac{dz}{z}. \tag{10.75}$$

## APPENDIX 10D

### GENERALIZED DERIVATION OF THE CHANGES IN POLE POSITIONS WITH RESPECT TO VARIATIONS IN THE FILTER COEFFICIENTS[15]

Consider the transfer function $H(z)$ given by:

$$H(z) = \frac{N(z)}{1 + \sum_{k=1}^{n} b_k z^{-k}} = \frac{N(z)}{\prod_{j=1}^{n} \left(1 - \frac{z^{-1}}{z^j}\right)} \tag{10.76}$$

where $z_j$ are the poles of $H(z)$ on the $z^{-1}$-plane, $b_k$ are the filter coefficients which specify the poles, $n$ is the number of poles and $N(z)$ is the transfer function of the zeros.

Equating the denominators of equation (10.76) we have

$$1 + \sum_{k=1}^{n} b_k z^{-k} = \prod_{j=1}^{n} \left(1 - \frac{z^{-1}}{z_j}\right). \tag{10.77}$$

Partially differentiating both sides with respect to $b_k$ yields:

$$z^{-k} = \frac{\partial}{\partial b_k} \prod_{j=1}^{n} \left(1 - \frac{z^{-1}}{z_j}\right) \tag{10.78}$$

$$z^{-k} = \frac{\partial}{\partial b_k} \left[\left(1 - \frac{z^{-1}}{z_1}\right) \cdot \left(1 - \frac{z^{-1}}{z_2}\right) \cdots \left(1 - \frac{z^{-1}}{z_n}\right)\right] \tag{10.79}$$

$$z^{-k} = \sum_{m=1}^{n} \frac{z^{-1}}{z_m^2} \cdot \frac{\partial z_m}{\partial b_k} \prod_{\substack{j=1 \\ j \neq m}}^{n} \left(1 - \frac{z^{-1}}{z_j}\right) \tag{10.80}$$

at the $m$th pole $z^{-1} = z_m$

$$\therefore \quad z_m^k = \frac{z_m}{z_m^2} \frac{\partial z_m}{\partial b_k} \prod_{\substack{j=1 \\ j \neq m}}^{n} \left(1 - \frac{z_m}{z_j}\right).$$

Hence

$$\frac{\partial z_m}{\partial b_k} = \frac{z_m^{k+1}}{\prod_{\substack{j=1 \\ j \neq m}}^{n} \left(1 - \frac{z_m}{z_j}\right)}. \tag{10.81}$$

## APPENDIX 10E

### ARITHMETIC ROUND-OFF ERRORS GENERATED WITHIN DIRECT AND CANONIC FORMS OF A FIRST-ORDER DIGITAL FILTER

**Example 5.** Consider a first-order filter whose transfer function is given by;

$$H(Z) = \frac{1 + \alpha z^{-1}}{1 - \beta z^{-1}}. \tag{10.82}$$

Realizations of the above transfer function in the direct and canonic forms are shown in Figures 10.23a and 10.23b, respectively.

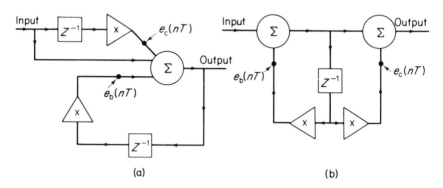

(a)

(b)

**Figure 10.23(a)** Direct realization    **Figure 10.23(b)** Canonic realization

$\alpha$ and $\beta$ are the coefficients.
$\beta$ should be less than unity.
$\alpha$ can take on any value.

#### (i) *Direct form realization*

The total average output noise power due to round-off noise generated within the filter can be obtained from equation (10.48) as:

$$\sigma_0^2 = \frac{2E_0^2}{12(2\pi j)} \oint \left[ \frac{1}{(1 - \beta z^{-1})} \frac{z^{-1}}{(z^{-1} - \beta)} \right] \frac{dz^{-1}}{z^{-1}} \tag{10.83}$$

where $1/(1 - \beta z^{-1})$ is the transfer function through which the two noise sources pass. The factor of 2 on the right-hand side of equation (10.83) shows that the noise is generated at two nodes within the filter.

The right-hand side of equation (10.83) is evaluated by calculating the residue due to the single pole at $z^{-1} = \beta$;

$$\sigma_0^2 = \frac{E_0^2 2\pi j}{12(2\pi j)} \times \frac{1}{1 - \beta^2} \tag{10.84}$$

hence the total average output noise power generated within the filter realized in direct form is:

$$\sigma_0^2 = \frac{E_0^2}{12} \frac{2}{1 - \beta^2}. \tag{10.85}$$

(ii) *Canonic realization*

The total average output noise power generated within the filter shown in Figure 10.23b is calculated using equation (10.51) as follows:

$$\sigma_0^2 = 1 \cdot \frac{E_0^2}{12(2\pi j)} \oint \left[ \frac{(1 - \alpha z^{-1})}{(1 - \beta z^{-1})} \times \frac{(z^{-1} + \alpha)}{(z^{-1} - \beta)} \right]_{z^{-1}} \frac{dz}{z} + 1 \cdot \frac{E_0^2}{12} \tag{10.86}$$

where the noise source $e_b(nT)$ passes through the transfer function

$$\frac{(1 + \alpha z^{-1})}{(1 - \beta z^{-1})}.$$

The other noise source $e_c(nT)$ is just added to the output, hence its contribution is given by $E_0^2/12$. Evaluation of equation (10.86) by the residue theorem gives

$$\sigma_0^2 = \frac{E_0^2}{12} \left[ \frac{(1 + \beta\alpha)(\beta + \alpha)}{\beta(1 - \beta\beta)} - \frac{\alpha}{\beta} \right] + \frac{E_0^2}{12} \tag{10.87}$$

which simplifies to:

$$\sigma_0^2 = \frac{2E_0^2}{12(1 - \beta^2)} + \frac{E_0^2(\alpha^2 + 2\alpha\beta - \beta^2)}{12(1 - \beta^2)}. \tag{10.88}$$

Comparison of equations (10.85) and (10.88) shows that the canonic form will generate less round-off noise if the term

$$\frac{E^2(\alpha^2 + 2\alpha\beta - \beta^2)}{12(1 - \beta 2)}$$

is negative. If it is positive, then the direct form is preferred. If the term is zero, then the direct and canonic forms produce the same amount of round-off noise.

### References

1. Gold B., and Rader, C., *Digital Processing Signals*, McGraw-Hill, 1969.
2. Jackson, L., 'An analysis of round-off noise in digital filters', *D.Sc. Thesis*, Stevens Institute of Tech., 1969.
3. Liu, B., 'Effects of finite wordlength on the accuracy of digital filters', *IEEE Transactions on Circuit Theory*, **CT-18** No. 6 (1971).
4. Lawrence, V. B., 'Use of orthogonal functions in the design of digital filters', *Ph.D. Thesis*, London University, Nov. 1972.
5. Bonzanigo, F., and Pellandini, F., 'Problems de realization des filters digitaux', *Agen*, July 1969.
6. Sandberg, I. W., 'Floating-point-round-off accumulation in digital-filter realization', *B.S.T.J.*, **46**, Oct. 1967.

7. Jackson, L., 'On the interaction of round-off noise and dynamic range in digital filters', *B.S.T.J.*, **49**, Feb. 1970.

8. Knowles, J. B., and Edwards, R., 'Effects of a finite-word-length computer in a sampled-data feedback system', *Proc. Inst. Elect. Eng.*, **112**, June 1965.

9. Liu, B., and Kaneko, T., 'Error analysis of digital filters realized with floating point arithmetic', *Proc. IEEE*, **57** No. 10, Oct. 1969.

10. Thomas, J. B., and Liu, B., 'Error problems in sampling representations', *IEEE Int. Conv. Rec.*, 1964.

11. Liu, B., and Thomas, J. B., 'Error problems in the reconstruction of signals from data', *Proc. Nat. Electronic Conf.*, **23**, Oct. 1967.

12. Bennett, W. R., 'Spectra of quantized signals, *B.S.T.J.*, **27**, July 1948.

13. Widrow, B., 'Statistical analysis of amplitude-quantized sampled data systems', *AIEE Trans. Appl. Ind.*, **79**, Jan. 1961.

14. Katzenelson, J., 'On errors introduced by combined sampling and quantization', *I.R.E. Trans. Automatic Control*, **AC-7**, April 1962.

15. Kaiser, J. F., 'Some practical considerations in the realization of linear digital filters', *Proc. 3rd Annual Alterton Conf.*, Oct. 1965.

16. Jackson, L., 'An analysis of limit cycles due to multiplication rounding in recursive digital filters', *Proc. 7th Annual Alterton Conf.*, 1969.

17. Bonzanigo, F., 'Constant-input behaviour of recursive digital filters', *IEEE Arden House Workshop on digital filtering*, New York, 1970.

18. Sandberg, I. W., and Kaiser, I. F., 'A bound limit cycles in fixed-point implementation of digital filters', *IEEE Trans. on Circuit Theory*, Nov. 1971.

19. Kan, E. K., and Aggarwal, J. K., 'Error analysis of digital filter employing floating point arithmetic', *IEEE Trans. on Circuit Theory*, **CT-18** No. 6, Nov. 1971.

20. Parker, S. R., and Hess, S. F., 'Limit cycle oscillations in digital filters', *IEEE Trans. on Circuit Theory*, **CT-18** No. 6, Nov. 1971.

21. Ebert, P. M., Mazo, J. E., and Taylor, M. G., 'Overflow oscillations in digital filters', *B.S.T.J.*, **48** No. 9, Nov. 1969.

22. Jackson, L. B., Kaiser, J. F., and McDonald, H. S., 'An approach to the implementation of digital filters', *IEEE Trans. on Audio and Electroacoustics*, **AU-16** No. 3, Sept. 1968.

23. Papoulis, A., *Probability, Random variables and Stochastic Processes*, McGraw-Hill, 1965.

24. Papoulis, A., *The Fourier Integral and its Applications*, McGraw-Hill, 1962.

25. Jury, E. I., *Theory and Application of Z Transform Method*, John Wiley, New York, 1964.

*Chapter 11*

# Optimization Techniques in Digital Filter Design

G. C. Bown

## 11.1 Introduction

The concept of optimization[1] in circuit design is more or less self-evident. We wish to adjust the design parameters (normally the component values) until the performance is optimum in some prescribed manner and in this way refine a given design into something more nearly ideal. To achieve this we must be able to assess the performance at any state of the design and this is usually achieved by sampling the response at $n$ values of the *independent variable* (e.g. frequency) and thus obtaining an error vector $\mathbf{E}$ where

$$\mathbf{E} = \begin{bmatrix} e_1 \\ e_2 \\ \cdot \\ \cdot \\ \cdot \\ e_n \end{bmatrix}. \tag{11.1}$$

The concept of an error vector may be clarified by a simple example that might arise in the study of non-recursive filters (Chapter 6). These have been studied extensively by optimization techniques[2,3]. The filter whose response to a unit pulse is $a_1, a_2, a_1, 0, 0 \ldots$ has a transfer function $H(\omega)$,

$$H(\omega) = (a_2 + 2a_1 \cos \omega T) \exp(-j\omega T)$$

in which the coefficients $a_1$ and $a_2$ are available for adjustment. To be specific and make the example as simple as possible, we will consider only the magnitude of $H(\omega)$, and rewrite the characteristic as $g(\theta)$:

$$g(\theta) = x_1 + x_2 \cos \theta \tag{11.2}$$

where $\theta$ is the independent variable and $x_1$, $x_2$ are coefficients to be chosen by the designer (i.e. the variables from the standpoint of our optimization problem). The ideal response can be given in terms of a target specification at $n$ discrete values of $\theta$ and for the purpose of the example let us assume that this specification is given by the following table

| $\theta$ | $g(\theta)$ |
|---|---|
| 0 | 1 |
| $\pi/2$ | 1 |
| $\pi$ | 0 |

The three elements of the error vector may now be written explicitly as

$$\left. \begin{array}{l} e_1 = x_1 + x_2 - 1 \\ e_2 = x_1 - 1 \\ e_3 = x_1 - x_2 \end{array} \right\}. \tag{11.3}$$

These give the differences between the values of $g(\theta)$ computed from equation (11.2) and the values specified in the table. The specification will be fully satisfied when each of these errors is zero and hence $\mathbf{E}$ is a null vector. By giving $x_1$, $x_2$ numerical values, corresponding values may be found for $\mathbf{E}$ as shown in the following table.

|       | Example 1 | Example 2 |
|-------|-----------|-----------|
| $x_1$ | 0         | 1         |
| $x_2$ | 0         | 1         |
| $e_1$ | $-1$      | 1         |
| $e_2$ | $-1$      | 0         |
| $e_3$ | 0         | 0         |

Two questions now arise: can we satisfy the specification completely and if not can we say which of two given designs is the better? With regard to the first point there are two degrees of freedom and three requirements to satisfy, so that we cannot hope to reduce each element of $\mathbf{E}$ to zero and in all that follows we shall assume that the number of sample points $n$ is greater than the number of variables $m$. This being so we are forced to consider what is the best solution and hence answer the second question.

Taking the results shown in the previous table we see that for the second example two of the three errors have been reduced to zero so that we seem closer to the ideal than in the first example where only one error was zero. On the other hand, we may consider that the result is only as good as the error at the worst sample point, since a realistic specification might well be drawn up on this basis. If this is so, then the second example is no better than the first, since the magnitude of the largest error is unity in both. This difficulty can only be resolved by basing our decision on a scalar quantity which in turn is a function of the vector $\mathbf{E}$. In the parlance of optimization this is termed the *objective function*.

The objective function is a scalar measure of the error existing between the performance actually attained and the specification. It should decrease as the performance is improved and be equal to zero for the ideal case where $\mathbf{E}$ is a null vector. The problem of optimizing the circuit design then becomes equivalent to the mathematical problem of minimizing a function of $m$ variables.

There are many possible ways of defining such a function. Two common ones are the sum of squares criterion:

$$\phi(x_1, x_2, \ldots, x_m) = (e_1^2 + e_2^2 \cdots + e_n^2)^{\frac{1}{2}} \tag{11.4}$$

and the maximum modulus criterion:

$$\phi(x_1, x_2, \ldots, x_m) = \max(|e_1|, |e_2|, \ldots, |e_n|). \tag{11.5}$$

The solutions corresponding to the minimum value for the two alternative criteria will be termed the least squares and minimax solutions, respectively. From a mathematical standpoint the least squares problem is attractive since it can be solved by the application of elementary principles of the calculus. The minimax problem, on the other hand, cannot be dealt with in this way since the first partial derivatives of $\phi$ with respect to the variables $x_1, x_2, \ldots, x_m$ are not continuous.

However, the minimax criterion is one which is often very attractive from the points of view of the specifier and user of a filter. Therefore, it has received considerable attention[2–6] and the non-recursive filter case has proven to be very amenable to solution by modified linear programming techniques as we see below. Figure 6.11 is an example of a bandpass filter designed with these techniques. Reference 2 gives extensive data for design of finite-duration impulse response bandpass filters.

Recursive filters have been designed by optimization techniques, using several criteria[6]. These filters usually result in non-linear relations between coefficients and performance measures. It has been found possible in some cases to carry out transformations which allow linear programming to be used.

### 11.2  The minimax criterion and the finite duration response

We stated above that the minimax problem has the property that the first partial derivations of $\phi$ with respect to $x_1, x_2, \ldots, x_m$ are not continuous. To clarify this, consider the example with $x_1 = 1$. Equation (11.3) then reduces to

$$\left. \begin{aligned} e_1 &= x_2 \\ e_2 &= 0 \\ e_3 &= 1 - x_2 \end{aligned} \right\} \tag{11.6}$$

If we now allow $x_2$ to vary and use the maximum modulus criterion, then $\phi$ will appear as shown in Figure 11.1a. The function certainly has a well-defined minimum at $x_2 = 0.5$ but the derivative is discontinuous. Another significant phenomenon may be observed by setting $x_1 = 0$. Equation (11.3) then reduces to

$$\left. \begin{aligned} e_1 &= x_2 - 1 \\ e_2 &= -1 \\ e_3 &= -x_2 \end{aligned} \right\} \tag{11.7}$$

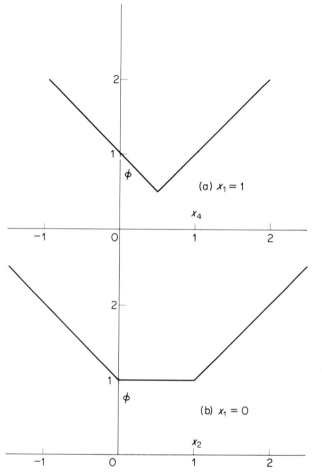

**Figure 11.1** Features of the maximum modulus criterion:
(a) $x_1 = 1$; (b) $x_1 = 0$.

and the form of $\phi$ is then as shown in Figure 11.1b. In this case we see that the minimum does not occur at a unique point but over the interval

$$0 \leqslant x_2 \leqslant 1$$

thus exhibiting another feature of functions with discontinuous derivatives.

It is now a logical step to remove the constraints on $x_1$ and consider $\phi$ to be a function of two variables. To show the result of this graphically we draw a contour diagram with each contour corresponding to a prescribed value of $\phi$ and obtain the result shown in Figure 11.2. It can be seen that the contours have sharp corners or cusps and can exhibit an abrupt change in shape. These abrupt changes are due to the transfer of $\phi$ from equality with one error to

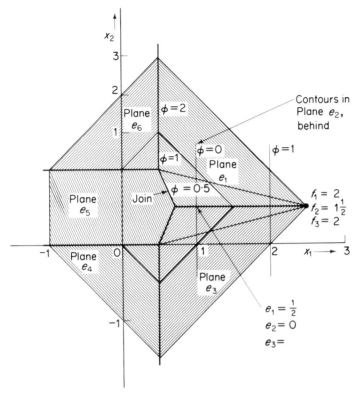

**Figure 11.2** Contour diagram

another as the variables are changed and can best be seen by redefining the objective function in the following way

$$\phi(x_1, x_2, \ldots, x_m) = \max(e_1, e_2, e_n, e_{n+1}, \ldots, e_{2n}) \tag{11.8}$$

where

$$e_{n+i} = -e_i. \tag{11.9}$$

This is equivalent to the original definition given by equation (11.5) but is in a more manageable form, and as such represents an important first step towards a solution. For the numerical example we now have six errors to consider as defined by

$$
\left.
\begin{aligned}
e_1 &= x_1 + x_2 - 1 \\
e_2 &= x_1 - 1 \\
e_3 &= x_1 - x_2 \\
e_4 &= -x_1 - x_2 + 1 \\
e_5 &= -x_1 + 1 \\
e_6 &= -x_1 + x_2
\end{aligned}
\right\} \tag{11.10}
$$

Consider now that these errors are measured along a third axis normal to the $x_1, x_2$ plane shown in Figure 11.2. They will now be represented by six planes (numbered as the $e$'s in equation 11.10) and $\phi$ will be defined by that surface which is the highest above the $x_1, x_2$ plane. This surface will therefore be that of a convex polyhedron with its lowest point, in general, at a vertex. This vertex results from the intersection of three planes and with reference to equation (11.10) it can be shown that these are numbers 1, 3 and 5. These three planes are active for values of $\phi$ up to and including $\phi = 0.5$ and give rise to three-sided contours as shown in Figure 11.2. For larger values of $\phi$, two more planes, numbers 4 and 6 become active giving five-sided contours. The sixth plane, number 2, is not active at any point and could be discarded without affecting the solution. Having determined graphically the three planes defining the vertex at the lowest point we obtain the solution

$$\left. \begin{array}{l} x_1 = 0.75 \\ x_2 = 0.5 \\ \phi = 0.25 \end{array} \right\}. \tag{11.11}$$

The graphical procedure, although very instructive, is tedious for two variables and virtually impossible for more than two variables, so that we must next derive a more formal and streamlined procedure. This is described in the following section but at this point it may assist the reader to point out that three principal steps are involved:

(a) formulation of the problem as a constrained optimization problem;
(b) sufficient conditions for a solution (i.e. minimum value of $\phi$);
(c) an algorithm by which a transition from a to b may be made in a finite number of steps.

One further point that should be clarified here is that in contrast to the traditional linear programming problem of operations research, we do not make the assumption that the variables $x_i$ are constrained to be positive. In what follows we will assume that these variables are unconstrained and may take positive or negative values.

## 11.3 Formulation and solution of the problem

We have already made a first step to a solution by augmenting the error vector as shown in equations (11.8) to (11.10). In matrix notation we may write this as

$$\mathbf{E}^* = \begin{bmatrix} \mathbf{E} \\ -\mathbf{E} \end{bmatrix} \tag{11.12}$$

where the vector $\mathbf{E}^*$ now has $2n$ elements. One further step, the addition of a new variable $\lambda$ to each error, will produce a constrained optimization problem. Applying this idea to equation (11.10) we obtain

$$y_1 = x_1 + x_2 + \lambda - 1$$
$$y_2 = x_1 + \lambda - 1$$
$$y_3 = x_1 - x_2 + \lambda$$
$$y_4 = -x_1 - x_2 + \lambda + 1 \qquad (11.13)$$
$$y_5 = -x_1 + \lambda + 1$$
$$y_6 = -x_1 + x_2 + \lambda$$

and it should be apparent that our problem can now be stated in the following way. Minimize $\lambda$ such that

$$y_1, y_2, \ldots, y_6 \geqslant 0. \qquad (11.14)$$

This is equivalent to the original problem since for given values of $x_1$, $x_2$ the minimum value of $\lambda$ consistent with the constraints (11.14) is equal to $\phi$. From this point on therefore, we may consider the new variable $\lambda$ as the objective function. In matrix notation these equations may be written in the form

$$\mathbf{Y} = \mathbf{E^*} + \lambda \mathbf{U} \qquad (11.15)$$

where $\mathbf{U}$ is a column vector with each element equal to unity. For practical computation the system of equations is best set out as an array of $2n$ rows and $(m + 2)$ columns as shown in Table 11.1.

Since we have a system of $2n$ equations in $(m + 1)$ independent variables, it follows that the objective function $\lambda$ can be expressed in the following form

$$\lambda = \sum a_i y_i + b \qquad (11.16)$$

where $i$ takes $(m + 1)$ discrete values in the range 1 to $2n$. If each coefficient $a_i$ is positive, then the minimum must occur for each $y_i$ equal to zero, since these are constrained variables. The corresponding value of $\lambda$ is equal to $b$. We finally must consider how we carry out the necessary algebraic manipulation of the system of equations shown in Table 11.1 [or more generally by equation (11.15)] in order to obtain $\lambda$ as a dependent variable in the form of equation (11.16). The first step is to obtain a so-called feasible solution [one satisfying the constraints (11.14)] and the second step is to obtain a sequence of feasible solutions such that $\lambda$ never increases, leading finally to the optimum solution.

We start with Table 11.1 and set each independent variable $x_1$, $x_2$ and $\lambda$ to zero. This fixes a point in 3-dimensional space, but one which is not feasible since the constraints on $y_1$ and $y_2$ are violated. This can always be remedied by allowing $\lambda$ to increase from zero until the constraints are just satisfied and in this case we require.

$$\lambda = 1$$

as shown at the foot of Table 11.1. One (or more) of the constrained variables will have been driven to zero by this move and a new point can be established by exchanging the appropriate constrained variable with the independent variable that has been changed from its initial value to zero. In this case we exchange $y_1$ with $\lambda$ and the element in the appropriate row and column is known as the pivot. The pivot elements have been underlined in the tables. The row in which the pivot lies is known as the pivot row and the column as the pivot column and an exchange step consists of solving the pivot row for the independent variable corresponding to the pivot column and substituting this in the remaining rows. In the example, this means that the first equation is solved for $\lambda$ and the result substituted in the remaining five equations. This leads to the new system of equations given in Table 11.2 which is an array of the same size in which $\lambda$ and $y_2$ have changed places. As before we consider that we are at the point defined by setting the independent variables $x_1$, $x_2$ and $y_1$ to zero. We now exchange the independent variables with the appropriate dependent variables starting with the first column and proceeding systematically in this way until the solution in the desired form is obtained, i.e. when all the independent variables are $y$ variables and all the coefficients in the $\lambda$ row are positive. We require one exchange to produce a feasible solution, a further $m$ exchanges to remove the $x$ variables from the set of independent variables and a further indeterminate number of exchanges to obtain all positive coefficients in the $\lambda$ row.

Returning to Table 11.2 and choosing the first column as the pivot column, we see that $x_1$ must undergo a positive change for $\lambda$ to decrease and the maximum permissible change in $x_1$ is set by $y_6$. Row 6 therefore becomes the pivot row and carrying out an exchange of $y_6$ and $x_1$ we obtain the results shown in Table 11.3. Two further exchanges, as shown by Tables 11.4 and 11.5 lead to the final solution. The formal rules for choosing a pivot element in column $j$ are as follows:

*Case 1.* The independent variable is an $x$ variable [as in the first $(m + 1)$ exchanges].

(a) Note the sign of the coefficient in the $\lambda$ row.
(b) Search the column $j$ over elements of that sign excluding rows corresponding to $x$ variables.
(c) If the sign were positive choose as pivot that element for which the ratio

$$a_{i,m+2}/a_{ij}$$

is a minimum.
(d) If the sign were negative choose as pivot that element for which the ratio

$$a_{i,m+2}/a_{ij}$$

is a maximum.

*Case 2.* The independent variable is a $y$ variable [as in exchanges after the first $(m + 1)$].

(a) Note the sign of the coefficient in the $\lambda$ row.

(b) If the sign were positive proceed to the next column, otherwise
(c) Choose as pivot that element for which the ratio

$$a_{i,m+2}/a_{ij}$$

is a maximum.

Having selected the pivot in this way we require some streamlined procedure for obtaining the new system of equations with the variables interchanged. This is provided by the exchange algorithm due to Stiefel[7].

## 11.4  The exchange algorithm

The system of equations is represented by a matrix of $2n$ rows and $(m + 2)$ columns as shown in the tables. If $a_{ij}$ and $b_{ij}$ are the elements of the old and new matrices respectively and if $a_{rs}$ is the pivot element then the algorithm is as follows:

$$b_{rj} = -a_{rj}/a_{rs} \qquad j \neq s \tag{11.17}$$

$$b_{ij} = a_{ij} - a_{is}a_{rj}/a_{rs} \qquad i \neq r, j \neq s \tag{11.18a}$$

$$b_{is} = a_{is}/a_{rs} \qquad i \neq r \tag{11.19}$$

$$b_{rs} = 1/a_{rs}. \tag{11.20}$$

By substituting equation (11.17) into (11.18a) we see that a somewhat more efficient computational procedure results if the latter equation is replaced by

$$b_{ij} = a_{ij} + a_{is}b_{rj} \qquad i \neq r, j \neq s. \tag{11.18b}$$

Note that, in words, the inequality $i \neq r$ means 'excluding the pivot row' and $j \neq s$ means 'excluding the pivot column'. The use of this algorithm may be clarified with the aid of a numerical example and for this purpose we will choose Table 11.4.

The pivot has already been selected (row 4, column 3) and we start by modifying the pivot row according to equation (11.17). This leads to the result shown in Table 11.6. The next step is to modify all elements not lying in the pivot row or pivot column according to equation (11.18b) and so obtain the result shown in Table 11.7. Modification of the pivot column then leads to Table 11.8 and finally replacing the pivot itself by its reciprocal leads to the result previously shown in Table 11.5 and so completes the exchange step.

## 11.5  Practical aspects

In the preceding notes we have considered the case where the errors are linearly dependent on the variables. In cases where this is not so the procedure must be used iteratively, employing a linear approximation for the errors at the start of each cycle. Some details of this procedure may be found in a paper by Ishizaki and Watanabe.[8]

The chief disadvantage of the minimax optimization procedure is the size of the $2n \times (m + 2)$ array and consequent high storage requirements imposed upon the computer in comparison with the least squares case. It is known that the solution is defined by the intersection of $(m + 1)$ hyperplanes, but unfortunately it is not generally known at the outset which are the particular errors corresponding to these and the entire set of $2n$ equations must be carried. If however, for a particular problem, information is available by which some of these errors may be discarded considerable savings in both computation time and storage can be effected.

### References

1. Bown, G. C. S., and Geiger, G. V., 'Design and optimisation of circuits by computer', *Proc. IEE*, **118** No. 5, 649–661 (1971).
2. Rabiner, L. R., Gold, B., and McGonegal, C. A., 'An approach to the approximation problem for nonrecursive digital filters', *IEEE Trans. on Audio and Electroacoustics*, **AU-18** No. 2, 83–106 (1970).
3. Rabiner, L. R., 'Linear program design of finite impulse response (FIR) digital filters', *IEEE Trans. on Audio and Electroacoustics*, **AU-20** No. 4, 280–288 (1972).
4. Rabiner, L. R., 'Techniques for designing finite-duration impulse-response digital filters', *IEEE Trans. on Communication Technology*, **COM-19** No. 2, 188–195 (1971).
5. Helms, H. D., 'Digital filters with equiripple or minimax responses', *IEEE Trans. on Audio and Electroacoustics*, **AU-19** No. 1, 87–94 (1971).
6. Thajchayapong, P., and Rayner, P. J. W., 'Recursive digital filter design by linear programming, *IEEE Trans. on Audio and Electroacoustics*, **AU-21** No. 2, 107–112 (1973).
7. Stiefel, E. L., *An Introduction to Numerical Mathematics*, Academic Press, 1963.
8. Ishizaki, T., and Wanatabe, H., 'An iterative Chebyshev approximation method for network design', *IEEE Trans. on Circuit Theory*, **CT-15** No. 4, 326–336 (1968).

Table 11.1

|       | $x_1$ | $x_2$ | $\lambda$ | 1   |
|-------|-------|-------|-----------|-----|
| $y_1$ | 1     | 1     | 1         | −1  |
| $y_2$ | 1     | 0     | 1         | −1  |
| $y_3$ | 1     | −1    | 1         | 0   |
| $y_4$ | −1    | −1    | 1         | 1   |
| $y_5$ | −1    | 0     | 1         | 1   |
| $y_6$ | −1    | 1     | 1         | 0   |

$\lambda = 1$.

Table 11.2

|       | $x_1$ | $x_2$ | $y_1$ | 1 |
|-------|-------|-------|-------|---|
| $\lambda$ | $-1$ | $-1$ | 1 | 1 |
| $y_2$ | 0 | $-1$ | 1 | 0 |
| $y_3$ | 0 | $-2$ | 1 | 1 |
| $y_4$ | $-2$ | $-2$ | 1 | 2 |
| $y_5$ | $-2$ | $-1$ | 1 | 2 |
| $y_6$ | $\underline{-2}$ | 0 | 1 | 1 |

$$x_1 = 0.5$$

Table 11.3

|       | $y_6$ | $x_2$ | $y_1$ | 1 |
|-------|-------|-------|-------|---|
| $\lambda$ | 0.5 | $-1$ | 0.5 | 0.5 |
| $y_2$ | 0 | $\underline{-1}$ | 1 | 0 |
| $y_3$ | 0 | $-2$ | 1 | 1 |
| $y_4$ | 1 | $-2$ | 0 | 1 |
| $y_5$ | 1 | $-1$ | 0 | 1 |
| $x_1$ | $-0.5$ | 0 | 0.5 | 0.5 |

$$x_2 = 0.$$

Table 11.4

|       | $y_6$ | $y_2$ | $y_1$ | 1 |
|-------|-------|-------|-------|---|
| $\lambda$ | 0.5 | 1 | $-0.5$ | 0.5 |
| $x_2$ | 0 | $-1$ | 1 | 0 |
| $y_3$ | 0 | 2 | $-1$ | 1 |
| $y_4$ | 1 | 2 | $\underline{-2}$ | 1 |
| $y_5$ | 1 | 1 | $-1$ | 1 |
| $x_1$ | $-0.5$ | 0 | 0.5 | 0.5 |

$$y_1 = 0.5.$$

Table 11.5

|       | $y_6$ | $y_2$ | $y_4$ | 1 |
|-------|-------|-------|-------|---|
| $\lambda$ | 0.25 | 0.5 | 0.25 | 0.25 |
| $x_2$ | 0.5 | 0 | $-0.5$ | 0.5 |
| $y_3$ | $-0.5$ | 1 | 0.5 | 0.5 |
| $y_1$ | 0.5 | 1 | $-0.5$ | 0.5 |
| $y_5$ | 0.5 | 0 | 0.5 | 0.5 |
| $x_1$ | $-0.25$ | 0.5 | $-0.25$ | 0.75 |

Solution: $x_1 = 0.75$, $x_2 = 0.5$.

Table 11.6

| | | | |
|---|---|---|---|
| 0·5 | 1 | −0·5 | 0·5 |
| 0 | −1 | 1 | 0 |
| 0 | 2 | −1 | 1 |
| 0·5 | 1 | −2 | 0·5 |
| 1 | 1 | −1 | 1 |
| −0·5 | 0 | 0·5 | 0·5 |

Table 11.7

| | | | |
|---|---|---|---|
| 0·25 | 0·5 | −0·5 | 0·25 |
| 0·5 | 0 | 1 | 0·5 |
| −0·5 | 1 | −1 | 0·5 |
| 0·5 | 1 | −2 | 0·5 |
| 0·5 | 0 | −1 | 0·5 |
| −0·25 | 0·5 | 0·5 | 0·75 |

Table 11.8

| | | | |
|---|---|---|---|
| 0·25 | 0·5 | 0·25 | 0·25 |
| 0·5 | 0 | −0·5 | 0·5 |
| −0·5 | 1 | 0·5 | 0·5 |
| 0·5 | 1 | −2 | 0·5 |
| 0·5 | 0 | 0·5 | 0·5 |
| −0·25 | 0·5 | −0·25 | 0·75 |

# Index